Glenn Brown

WATER-CLOSETS

A

HISTORICAL, MECHANICAL,

AND

SANITARY TREATISE

Elibron Classics
www.elibron.com

Elibron Classics series.

ISBN 1-4021-6630-3 (paperback)
ISBN 1-4212-8179-1 (hardcover)

This Elibron Classics Replica Edition is an unabridged facsimile of the edition published in 1884 by the Industrial Publication Company, New York.

WATER-CLOSETS.

A

HISTORICAL, MECHANICAL,

AND

SANITARY TREATISE.

BY

GLENN BROWN,

ARCHITECT. ASSOCIATE AMERICAN INSTITUTE OF ARCHITECTS.

NEW YORK:
THE INDUSTRIAL PUBLICATION COMPANY.
1884.

PREFACE.

The principal part of this book is a republication of the series on WATER-CLOSETS which appeared in the *American Architect and Building News* during the year 1883.

The matter has been revised, and the section on tanks and supply-valves added.

All illustrations were engraved from drawings made by the author, in a uniform style, specially for this work.

In book form the author hopes that the matter contained in it may be of service to all (and I think there are many such) who desire information on this plumbing fixture either from a mechanical, historical, or *sanitary* standpoint. The information contained herein has been gleaned from all available sources.

GLENN BROWN.

607 Louisiana Avenue,
Washington, D. C., 1884.

CONTENTS.

WATER-CLOSETS.

SECTION I.

ANCIENT CLOSETS.

WATER-CLOSETS as conveniences are so necessary, and in their proper construction are so important to our comfort and health, that I think any research into their history will repay us by increasing our knowledge on the subject.

It will be necessary in the first place to review, cursorily, other modes of convenience in countries where, and at times when, water-closets properly speaking were not in use.

Naturally, we first examine the ruins of Egypt, where existed the earliest traces of civilization, for remains of water-closets. The small private and detached rooms which we find in the remains of Egyptian houses were probably used as privies. Ewbank, in his work on hydraulics, calls the summer chamber* of Eglon, king of Moab, a water-closet.

If I thought proper to follow the example of Ewbank I would call water-closets the private rooms which were in an isolated position in one of the halls, being near a door communicating with the other chambers. Sir J. Gardner Wilkinson says: "These rooms bear a striking resemblance to the before-mentioned private room of Eglon."†

That the Greeks‡ made use of privies in their houses is proved

---

* *Judges*, iii, 20–25

† *Egypt*, Sir J. Gardner Wilkinson, Vol. II., page 104.

‡ Viollet-le-Duc has given the Egyptians', Greeks' and Romans' privies (*latrines*) in their houses in the *Habitations of Man*, pages 92–102, 197 and 231.

by an old writer* from a passage in Aristophanes (Ecclesiaz., verse 1050). This was about four hundred years before Christ.

The Romans, if not the first in art, the leaders in all that pertained to luxury and comfort, were the first, as far as we can ascertain, to use water-closets. In Rome we find four kinds of receptacles for excreta. Close stools (*lasana*), in which the rich ancients sometimes used gold or silver bowls; vases (*gastra*) which were stationed on the roadways; public privies (*cloacina*), of which Sir William Gell tells us there were one hundred and forty-four in Rome; privies (*latrina*), probably for private use.

From their derivation we would infer the two classes last mentioned to be water-closets, *cloacina* being derived from *cloaca*, a sewer or drain, and *latrina* being a diminutive of *lavatrina*, a wash-bowl or basin. No doubt, as at the present day, the name included the room as well as the basin or receptacle contained in the room. Pompeii was covered with ashes and lava by an eruption of Mount Vesuvius more than eighteen hundred years ago, and among its ruins were found the most perfect remains of an ancient water-closet. Pompeii being a small town and a province of Rome, I would suppose the water-closets to have been more numerous and more elaborately ornamented in the capital city.

In the *Pompeiana* Sir William Gell,† describing a privy, writes of an arched recess about three feet deep discovered in the kitchen of one of the ruined dwellings of Pompeii, most inconveniently placed, according to American ideas (Fig. 1). The wood-work was gone when discovered, about 1819, but the marks of the hinges and fastenings were still visible. Three or four feet to one side were the remains of brick ovens. It would appear, according to a quotation from Pretonius, that in ancient Italy it was considered desirable to have in close proximity the place for the preparation of the food and

Fig. 1.—Privy in kitchen in Pompeii.
a, Niche. *b*, Ovens.

* *Mem de l' Academie des Sciences Inscriptions nouvellement etablie a Troyes en Champagne*. A. Troyetset, Paris. 1756.

† *Pompeiana*, Sir William Gell, Vol. II., page 171; illustration, page 151.

the receptacle for the excreta. The privies in or adjoining the kitchen were probably for the women, while others which have been discovered at the back of the house were for the men. These chambers were sometimes finished beautifully, having tessellated pavements and in some instances windows looking into small interior courts, an example which the moderns seem injudiciously prone to follow.

Fosbroke says, under the head of ancient water-closets,* "That of the Palace of the Cæsars is adorned with marble and mosaick. At the back of one is a cistern, the water of which is distributed by cocks to the different seats. The pipe and bason of one still remains at Pompeii and is like ours." (1825 A. D.)

In a paper read before the Society of Antiquaries in 1775, relating to discoveries in Pompeii, Sir William Hamilton† says: "Close to the Temple of Isis is a theatre, no more of which has been cleared than the scene and corridor which leads to the seats. In the corridor was a retiring place for necessary occasions, where the pipe to convey the water, and the bason like that of our water-closets (1775) still remain, the wood of the seat only having mouldered away by time."

The above-mentioned remains, seen by Sir William Hamilton, conclusively prove the existence of water-closets more than eighteen hundred years ago, with basin, water-supply and waste for the water, and excreta must necessarily have been carried off by a drain of some kind.

In all descriptions of ancient baths we find a latrina, that near the entrance of the ruins of the one in Pompeii, mentioned and shown by illustrations, but it is in no case fully described.‡

According to Olympiodorous there were in the Thermæ of Antoninus (about 200 A. D.) sixteen hundred seats of marble with holes like a close stool for the convenience of those who attended either the baths or the gymnasium.

F. Liger,|| in his work on this subject, informs us that the remains

* *Encyclopedia of Antiquities*, Rev. Thomas Dudley Fosbroke. Page 348. 1825.

† *Archæologia*, Vol. IV., page 168. 1777.

‡ *Encyclopedia Britannica*, Vol. III., page 435.

|| "*Fosses d'aissance Latrines Urinoirs et Vidanges.*" F. Liger, Architect.

of three privies were found in the ruins of Acteon's house, Pompeii.

One was situated against the wall on the alley, another under the steps, and the third was in the kitchen. The first was the only one that received light from the outside. Remains of privies (*latrines*) are still found in Lucretius's house, placed in a narrow closet.

A drawing by Piranesi represents a water-closet (*sterquilinium*) from the same town. There are three compartments placed in a large chamber, one of which has a seat; the other seems intended for use after the manner of Orientals (by squatting), while the third was evidently used as a urinal. The water is admitted by a pipe which runs through the wall; thence it flows in front of the seats through a gutter, falling into the urinal from a higher level, where it turns and runs under the closet seats, carrying away fæcal and other excrementitious matter. This was probably a cleanly arrangement, when we take into consideration the fact that the Romans cleaned off all sediment or other particles which adhered to the surface with a sponge or mop fixed on the end of a stick* (Fig. 2). In the work entitled

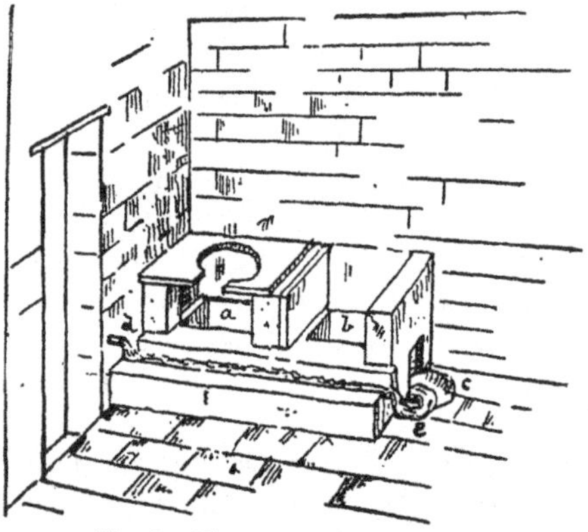

Fig. 2.—Water-Closet in Pompeii.
a. Seat for sitting on. *b.* Place for crouching. *c.* Urinal. *d.* Water-pipe. *e.* Stream running in urinal.

* Mazois' *Pompeia*.

[*Inscription found on a wall in a privy in Pompeii.*]

Quaeras censeo, si léges laboras
Nigri Tornicis etrium poetam
Qui carbone rudi patrique crita
Scribit carmina quae legunt cacantes.

"*Le Case et i Monumenti*" are mentioned remains of privies which were found in the house of Marco Lucretio. In these rooms remains or indications of tiling were found, and obscene figures pointedly drawn. Cochin and Belicard, in "*Les Antiquités d' Herculaneum*," mention and illustrate several seats with holes which were found in the palace of Serapis (1750), Puozzoli (Figs. 3 and 4).

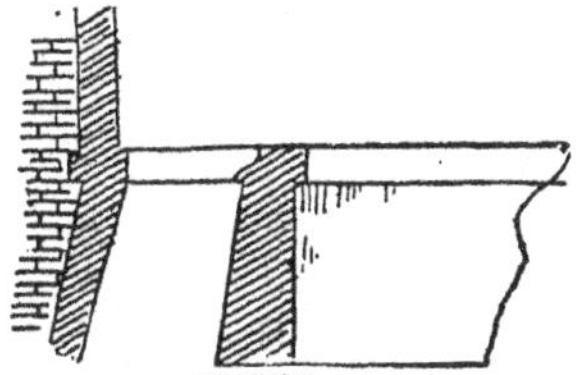

SECTION.

Fig. 3.

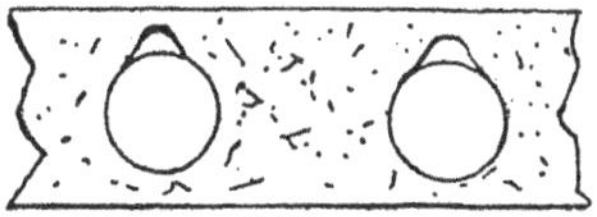

PLAN.

Fig. 4.

Privy from Palace of Serapis, Puozzoli.

Public latrines among the Romans had no seats, they being in the habit of crouching, after the manner of Orientals. In private houses marble seats were usually used; but in some cases the choice

Fig. 5.

Fig. 6.

Roman Seats in Louvre Museum.

of arrangements was given them, both kinds, with and without seats, being placed in the same room. In Constantine's time

(300 A. D.) and probably at an earlier date, seats in the shape of chairs, with backs and arms and elaborately carved legs and feet, were used for this purpose. A fine example was in the Louvre some years ago, which is supposed to have dated back to Constantine's reign (Figs. 5 and 6).

Vitruvius, strange to say, does not mention water-closets or privies in his description of private houses. Both ancient and modern writers seem to avoid the subject. Ewbank states that a number of water-closets were erected near the mosques and temples in Old Rome, Smyrna, Constantinople, and probably all ancient cities. He does not give his authorities for the statement, and it would appear as if it were a little overdrawn.* With the decline of the Roman empire, water-closets, following her other luxuries and comforts, seem to have gone out of date.

Giving a cursory review of the privies used in the middle ages we leave unaccounted for nearly a thousand years, during which time there seems to have been no privies used inside houses. The castle of Rochester, designed in part by Bishop Gundulph about 1088, has outlets for privies and sinks similar to the outlets for smoke, in the first instance going downward, instead of upward as in the case of their chimneys.

In a paper read by Edward King† before the Society of Antiquaries (1782 A. D.) in relation to a Saxon castle, he says: "In one of the corners is a narrow passage to a small closet in the wall, which served for a privy, having the usual kind of an outlet through a loop." (Fig. 7). Diagonally across from the privy is a similar passage and closet supposed to have been a well. This tower or castle was undoubtedly Saxon, having been given by William the Conqueror to William de Peverel and called the Castle on the Peake. We learn from Viollet-le-Duc‡ that

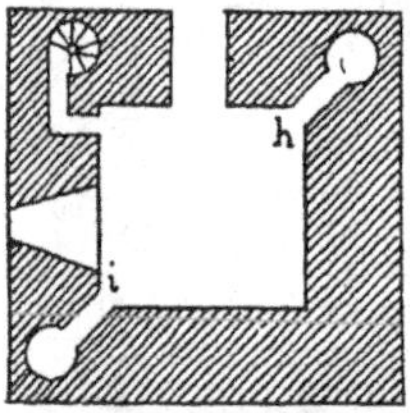

Fig. 7.—Plan of Castle on the Peake.
h, Privy. i, Well.

* *Hydraulics*, Thomas Ewbank, pages 261, 562. New edition, 1876.

† *Archæologia*, Vol. VI., page 242.

‡ *Dictionaire d' Architecture*, XIIth au XVIth, siècle, Vol. VI., pages 163–170.

castles (*châteaux*) at the commencement of the thirteenth century had privies (*latrines*) on each floor, constructed so as to avoid the inconveniences attached to these necessities. They were generally found projecting on corbels, from an angle formed by a buttress and the main wall (Figs. 8 and 9), so the fæcal matter might be cast well out and down the precipice into the woods which usually surrounded these buildings.

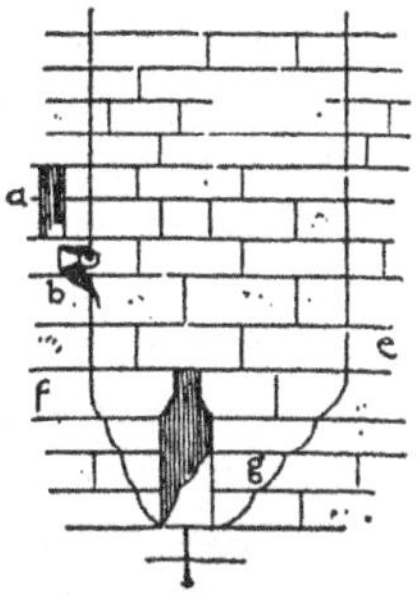

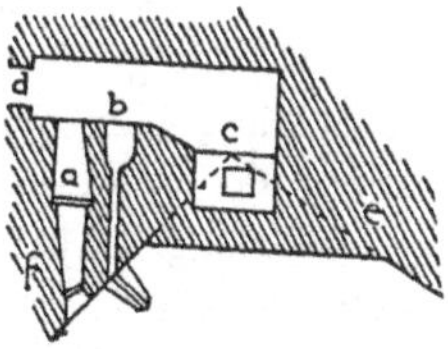

Fig. 8.—Elevation.
Fig. 9.—Plan.

Thirteenth-Century Privy.

a, Window. *b*, Urinal. *c*, Seat. *f*, Wall. *e* Buttress. *d*, Door. *g*, Corbels.

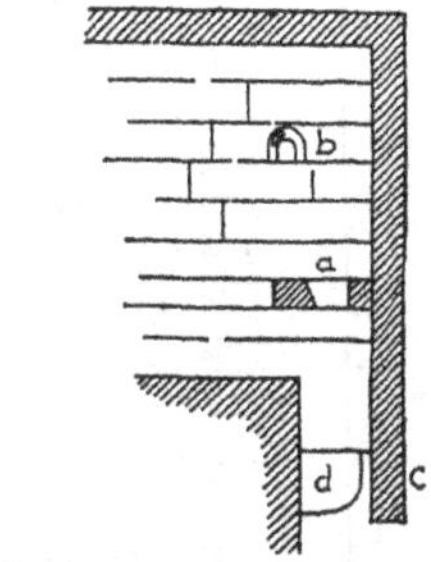

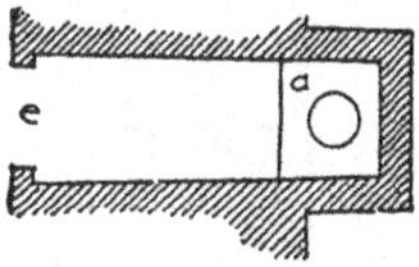

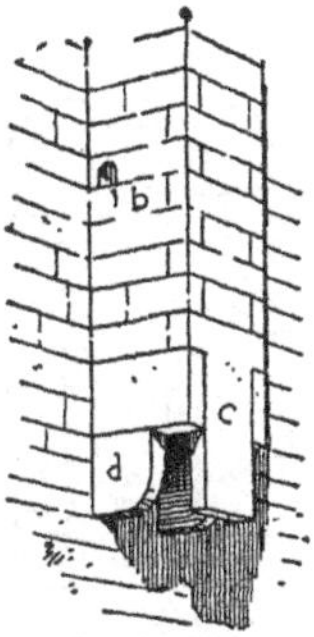

Fig. 10.—Section.
Fig. 11.—Plan.

Fig. 12.—Perspective Sketch.

Landsberg Latrine.

a, Seat. *b*, Window. *c*, Hanging T-shaped stone. *d*, Corbels. *e*, Door.

He describes one which is corbelled from the wall and buttress in the usual way. The seat is concealed and protected from missiles by the corbels, and it communicates with the halls and stairways by a door. In the same closet there is a urinal with drain, and a drip on the outside, and also a window opening into the outer air. In the castle of Landsberg on the Lower Rhine, which dates from the twelfth century, there is one (Figs. 10, 11 and 12) that differs from the above. The seat is guarded by a rectangular projection from the wall which is supported on corbels. The seat is protected from arrows and other projectiles by a stone shaped like a T, which hangs on and below the corbels. In garrisoned castles they always had projections reserved for privies, and they were separated from the main building by small passage-ways. There are privies of this kind in the castle of Chauvigny (Poitou), while the castle of Langley, Northumberland, England, has them, almost monumental in character or design. The castle of Marcousis has an extension devoted to latrines, a set of four for each story (Figs. 13 and 14). There is a passage-way extending from the main building with solid stone balustrade to prevent people from falling. There is a chute or drain from each seat leading to a double-arched cesspool. This part of the building is ventilated by a large window above the highest tier of latrines. The castle of Pierrefonds was constructed in the fourteenth century. It has a set of latrines arranged for each story in a semi-circular projection (Figs. 15, 16, and 17). The cesspool, as we will call the place for the receptacle of the excreta,

Fig. 13.—Elevation. Fig. 14.—Section.
Latrines, Marcousis Castle.
a, Seat. *b*, Chute or Drain. *c*, Entrance from Main Building. *d*, Window. *f*, Cesspool. *e*, Balustrade.

has a door for cleaning out the filth, and a window for ventilation. In the centre of this chamber is a large piece of cut stone on which a man can stand conveniently while cleaning out the pit. The chute or drain leads directly from the seats to the cesspool, and is carried through the roof for the purpose of ventilation; the first instance probably, of a soil-pipe being carried through the roof. These closets, and it

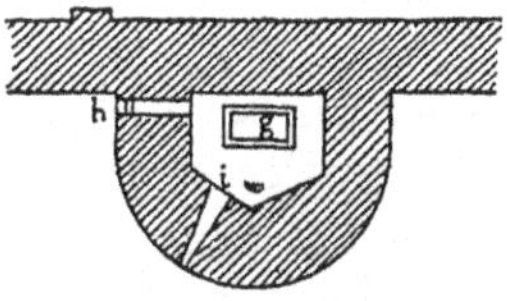

Fig. 15.—Plan of Cesspool.
h, Door. *i*, Window. *g*, Stone.

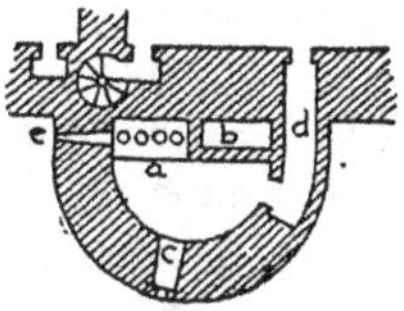

Fig. 16.—Plan of Principal Story.

Fig. 17.—Perspective Sketch.

Latrine Pierrefonds.

a, Seats. *b*. Chute or Drain. *e*, Vent. *c*, Window. *d*, Small Entrance Hall.

was nearly always the case at this period, as we have mentioned before, were separated from the main building by a short hall with a door at each end. This idea, in some instances, has been utilized in modern times. These closets also have windows and vents in each story.

Sometimes latrines were directly adjoining dungeons with drain running to the cesspool (Fig. 18).

Sight-seers are frequently shown through these latrines, and are informed that the cesspools are dungeons, the agonies which have been endured by prisoners in their depths being graphically depicted by the guides.

Curiously enough the first mention we have of a water-closet, properly speaking, after the ones used by the Romans, is connected with the publication of a poem by Sir John Harrington on the "Metamorphosis of Ajax" (a close stool or commode), published in 1596. This poem was occasioned, as Henry Harrington reports, by the author having invented a kind of water-closet for his house at Kelson.*

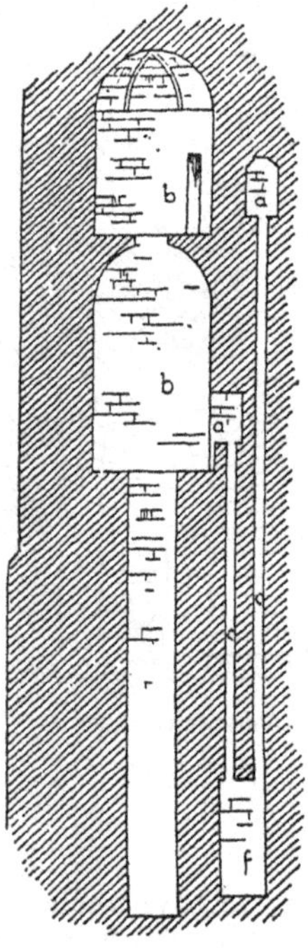

Fig. 18.
a, Latrine. *b*, Dungeon.
c, Chute or Drain.
f, Cesspool.

In the East water-closets, or what may be called water-closets, seem to have been used at an early date.

Ogilby, in a description of the city of Fez, says: "The river Fez subdivides into many clear-running channels through the streets, serving not only each private house, but churches, inns, hospitals, and other public places, to their great convenience. Round about the mosques are one hundred and fifty common places of ease, built four-square and divided into single stool-rooms, each furnished with a cock and marble cistern, which scoureth and keepeth all neat and clean, as if these places were intended for some sweeter employment."†

Tavernier, in his "Relations to the Seraglio," says: "The places for the easing of nature are on the right hand divided into four little rooms, which are always kept very clean, and paved with square pieces of white

* *Nugæ Antiquæ*, page xi. 1804. New edition.
† Ogilby's *Africa*, page 188. 1670.

marble. The Turks do not sit down as we do when we are in those places, but they squat down over the holes, which is not half a foot, or a little more, above the ground. The hole is covered with an iron plate, which rises and falls by a spring, and turning one way or the other, at the falling of the least weight upon it, resumes the station it was in before as soon as the ordure has fallen from it. I have observed that the Turks and generally all the Mohammedans do not use paper in such cases, or upon any indecent occasion, and when they go to those kind of places they carry along with them a pot of water, to wash themselves withal, and the iron plate is also made clean at the same time, and so the hole being always covered, and the iron plate always kept clean, there can be no ill smell in the place, and that the rather for the small current of water that passes under that place carries away all the ordure."

...."A little gallery which lies to the left hand of the bath, leads to places designed for the easing of nature, and every seat has a little cock which supplies them with water to wash themselves after they are done."

Baldwin Latham, quoting from Ewbank, would lead us to suppose the above-mentioned cocks were primarily for the purpose of washing the bowl of the closet, if there was one, while it was for the purpose of washing the person using the closet.

We see no mention of water-closets in Europe again until 1770. During the intervening period, while art had professedly entered into a new life, the useful arts seem to have been ignored.

In England a close stool or commode, called an ajax, or jakes, was generally used. According to Viollet-le-Duc and other writers, cleanliness, the first principle of sanitation, was thought of very little importance during this period.

Privies were used in Germany before they were in France. In Paris people were allowed to throw their excreta from the windows into the street, provided they gave the verbal warning, "*Gare l' eau*" three times. In Edinburgh the same custom prevailed; parties walking the streets had necessarily to cry "Haud your han," for fear of what might befall them.

The above practice was forbidden in Paris in 1395, but seems to have continued in Edinburgh until 1750. It was during this period also, that men walked the streets of Edinburgh carrying pails or buckets for the public convenience, and wearing a cloak large enough to envelope their customers and apparatus, crying "Wha wants me for a bawbee."

The Parisian code of laws, improved in 1513, expressly ordered that every house should have a privy, but as late as 1700 A. D. the police were instructed to close all houses whose occupants did not provide one within a month.

During the reign of Louis XIV. of France the courtiers found it necessary to ease themselves in the halls of the palace of Versailles because there were neither privies nor water-closets. Viollet-le-Duc tells us that close stools were used in the palace of Versailles in the eighteenth century. He also writes of an old lady belonging to the court of Louis the XV. of France, who passed through the halls of St. Cloud with him when a young man. The odor of an offensive vessel caused her to exclaim with pleasure, "This smell recalls the good old times." Some of our modern closets would have delighted the old lady.

Privies were not used in Spain until about 1770, and have not been used in Warsaw more than fifty years. As late as 1846 we find in Berlin "small closets located on the landings of the stairs, which require to be emptied every other night, to the no great satisfaction of the olfactory nerves."

Aubrey, writing in 1718, describes a water-closet which he had seen. He says: "Here (at Sir Francis Carew's, Beddington, Sur rey) I saw a pretty machine to cleanse an House of Office—viz., by a small stream of water no bigger than one's fingers, which ran into an engine made like a bit of a fire shovel which hung upon its centre of gravity, so that when it was full a considerable body of water fell down with some force and washed away the filth."

This must have been very much like the one seen in Fez by Ogilby.

S. S. Hellyer gives a cut of a water-closet taken from Osterly House, which he considers a type of the closets used in England one hundred to one hundred and fifty years ago. Describing it he says:* "A niche in a fair-sized room was formed to receive the marble closet-pan [bowl], and a door shutting up close to the seat hid the whole arrangement from sight. A lead soil-pipe was connected with the outlet-plug (plunger) waste of the pan, and continued from it to the drain, which was brought into the house to receive it. The soil-pipe had no ventilation." (Fig. 19).

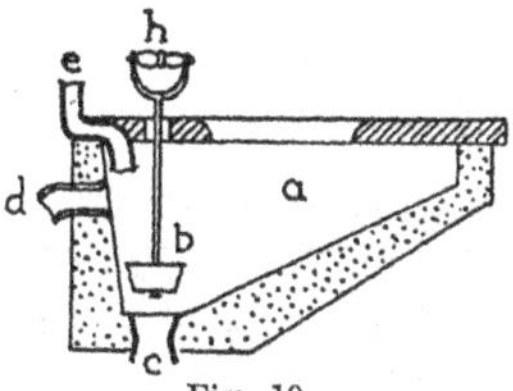

Fig. 19.
a, Bowl. *b*. Plunger. *c* Waste. *d*, Overflow. *e*, Supply. *h*, Handle.

A. J. Roubo describes a water-closet, and also gives several views to illustrate the subject in the great French book on joiners' work (L'Art Menuisier), published in

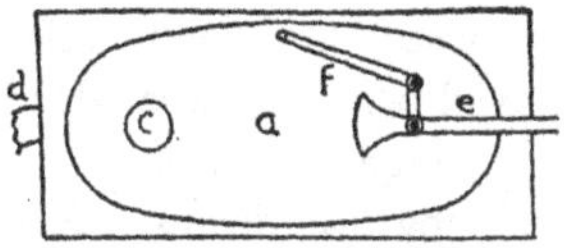

Fig. 20.—Plan, or Top View.
a, Bowl. *c*, Waste with plunger removed. *e*, Supply. *f*. Bidet-attachment. *d*, Overflow.

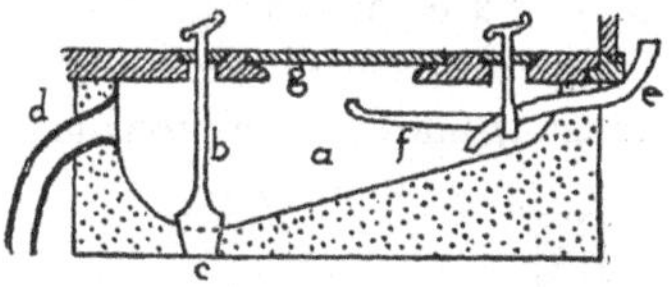

Fig. 21.—Longitudinal Section.
a, *c*, *d*, *e*, *f*, Same as Fig. 20. *b*, Plunger. *g*, Seat Cover.

1770.† I think these are the earliest illustrations of a water-closet extant. (Figs. 20–25). The bowl is cut from a single block

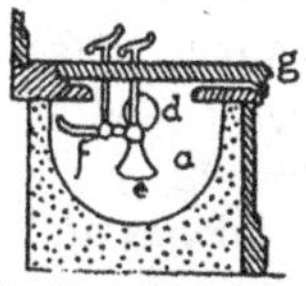

Fig. 22.—Cross-Section.
Letters same as above.

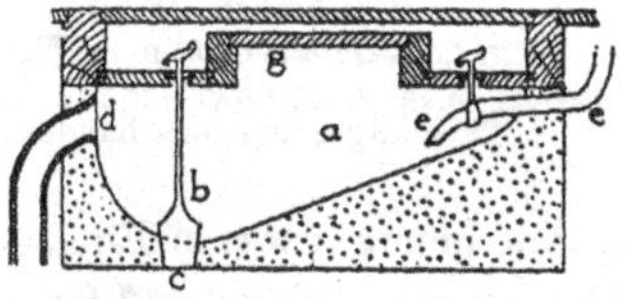

Fig. 23.—Section of Closet.
Without bidet-attachment and different arrangement of seat. Letters same as Fig. 21.

of stone, all the inside lines being curves so it will not retain the filth as readily, and be easier to clean than if it were in another

* *Sanitary Plumbing*. 1882.

† *L'Art Menuisier*, page 203, plate 69, Vol. II. 1770.

shape. The lead supply-pipe has T-shaped handles and a nozzle spread into the form of a fan, which is evidently for the purpose of spreading the water over a larger surface of the bowl than would be the case if the nozzle was of the same shape as the pipe. The above-mentioned closet is one of the simplest forms of plunger-closets, and if we are determined to use a closet of this class, then the simpler style would be better and more cleanly than those which have concealed and foul compartments in which the plunger works, and of which only experts have cognizance. In this closet we see a branch from the supply-pipe which could have been used only as a bidet-attachment. This has a hinge-joint so it can be folded back under the seat when not in use. The wood-work is carefully mitred, being tongued and grooved wherever it is advantageous, as is shown by the details. A lid covers the seat, and is made so it will fit as nearly air-tight as possible when it is closed. The closet described was called "a closet with a plug" (*Lieux de soupapes*), and it was in use in France a long time before it was in England, as the author informs us.

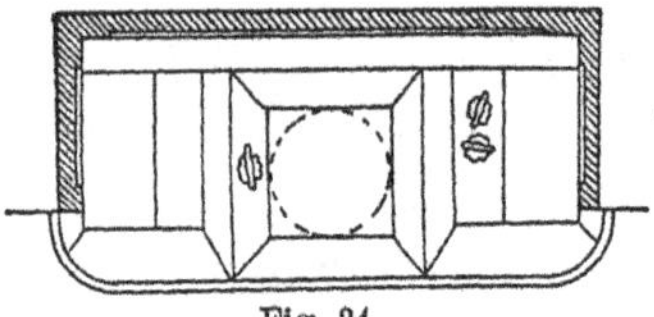

Fig. 24.
View of seat showing mitres and joints.

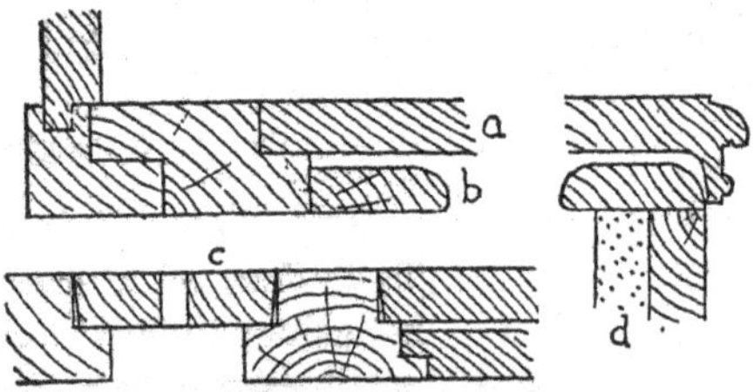

Fig. 25.—Details of Joiners' Work.
a, Cover. *b*, Seat. *d*, Bowl (stone). *c*, Opening for plunger and cock handles.

The first English patent* for a water-closet was issued in 1775 A. D., to Alexander Cummings, a watch-maker in Bond Street. This closet has a sliding valve, or slider, as it is called, between the trap and bowl. The title being "a water-closet upon a new construction," proves the use, we might say the common use, of water-closets during and before this time, but of a different pattern

* *British Patent-Reports*, Vol. XIV., No. 1105.

from Cummings's closet. Here we find the first recorded instance of a siphon trap being used under and in connection with a water-closet, although we have evidence of siphons having been used in the time of the Egyptians. Below are quoted a few extracts from Cummings's specifications: "The advantages of the said water-closet depend upon the shape of the pan or bason [bowl], the manner of admitting water into it [near the bottom], and on having the stink-trap hitherto used for water-closets so constructed that its contents shall or may be totally emptied every time the closet is used.... The stink-trap hitherto used for water-closets is too well known to require a description here; and although it may serve effectually to cut off all communication of smell from the drains, pipe and cesspool, it becomes in

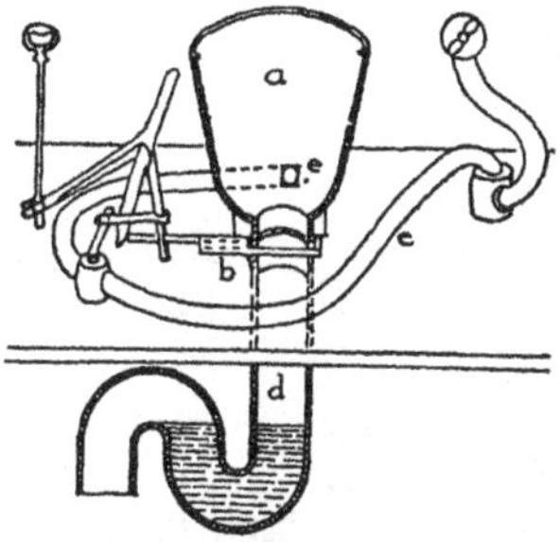

Fig. 26.—Cummings's Closet.
a, Bowl. *b*, Slider, or valve. *d*, Siphon-trap.

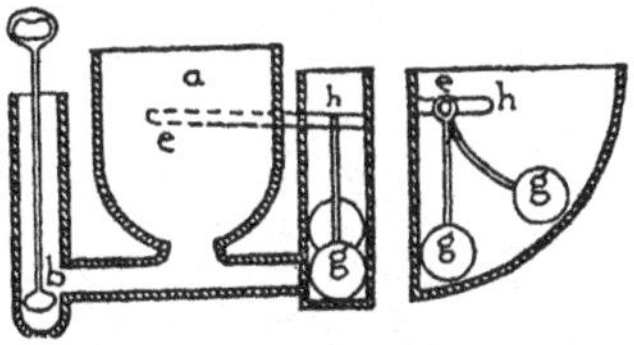

Fig. 27.—Prosser's Closet.
a, Bowl. *b*, Plunger. *h*, Chamber for ball-cock. *g*, Balls for cock. *e*, Supply.

itself a magazine of fetid matter, which emits an offensive smell every time that it is disturbed by using the closet. In this water-closet the pipe which carries off the soil and water is recurved about twelve or eighteen inches below the pan or bason [bowl] so as constantly to retain a quantity of water sufficient to cut off all communication of smell from below, and this stagnated water in the recurved part of the pipe is totally emptied and succeeded by fresh every time the pan or bason is emptied.... The horizontal sections are circular, or nearly so." The mechanism is very clearly shown in the drawing which accompanies the specification. If the siphon-trap fulfils the the duties specified, and we think it does, why complicate matters with the "slider."? (Fig. 26).

The next patent* was issued in 1777 to Samuel Prosser, a plumber living in the parish of Martin-in-the-fields, for a plunger-closet with a double-ball-cock for shutting off the water when it had risen to a certain height in the bowl, and for turning on the water-supply when it had fallen below a definite point. Here we see, for the first time, the (as we consider it) pernicious practice of placing the plunger and ball-cocks in compartments separated from the bowl, as far as the sight can penetrate, but connected as far as the deposits from the water and filth are concerned. This closet is described in the specifications as "a water-closet upon an entirely new construction, which will always remain free from any offensive smell." We, at the present time, know too well the folly

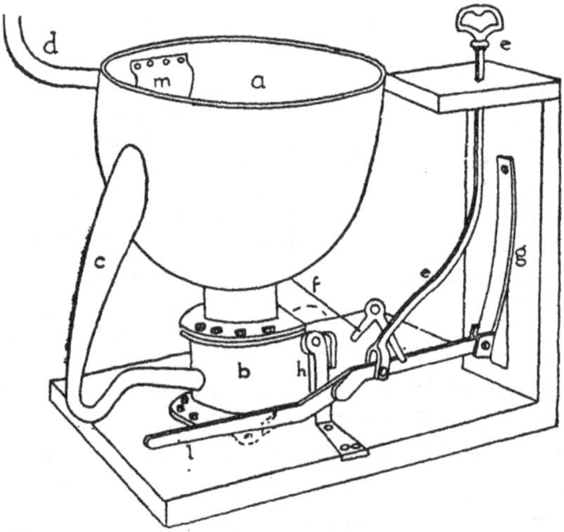

Fig. 28.—Bramah Closet.
a, Bowl. *b*, Valve-box. *c*, Overflow. *d*, Supply. *e*, Handle and pull-rod. *f*, Wire to tank. *g*, Spring. *h*, Valve-crank. *l*, Lever. *m*, Fan.

of such claims. Wherever concealed chambers exist connected directly with the bowl, there filth and decayed matter will accumulate, and will become not only offensive to our olfactory nerves, but hurtful to our health. The excreta, in finely-divided particles, in

* *British Patent Reports*, Vol. XV., No. 1160.

this and all similar cases, has free access to compartments that can never be cleaned without the assistance of an expert mechanic (Fig. 27).

In 1778 Joseph Bramah had a patent* issued to him for a water-closet with a valve at the bottom of the bowl, working on a hinge. This valve was for the purpose of cutting off all connection with the soil-pipe (Figs. 28, 29). It will be seen at a glance, that this closet was the forerunner of a large number of inventions, differing more or less from the original, of which the parent is certainly equal

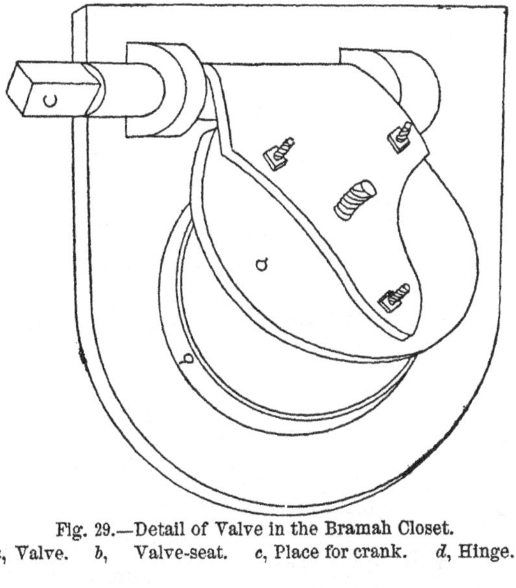

Fig. 29.—Detail of Valve in the Bramah Closet.
a, Valve. *b*, Valve-seat. *c*, Place for crank. *d*, Hinge.

to, if not superior to, any of its offspring. All valve-closets with a horizontal valve working on a hinge, and the numerous patterns of pan-closets, are an outgrowth from the Bramah. The valve works in a small chamber not much larger than the pipe. The mechanism is comparatively simple, and it has a simple U-trap as an overflow. We also note the flap or fan placed over the supply-pipe

* *British Patent-Reports*, Vol. XV., No. 1177.

for spreading the water over the bowl. I quote the following from the specifications: "In this closet two valves are simultaneously acted on by the handle. The one valve is placed under the bottom of the basin [bowl], and when closed retains any water that may be therein, thus cutting off all smell that may be therein. It moves on a horizontal axis, and in such a direction that its surface is the more thoroughly washed every time the contents of the basin is discharged, both by the water falling upon it whilst opening, and by its meeting the water when shutting to again. The other valve is placed in the service-box, or the top part of the feeding-pipe; this valve opens and shuts in a similar manner: always meeting the natural currrency of the water, when shutting, and falls from it when opening, and is not liable at any time to have its motion obstructed by the frost, no part of it being immersed in the water. Its movement is communicated through a lead-pipe or tube, soldered into the part where it acts, and is continued above the highest surface of the water in the reservoir." From the drawings it would appear that the bowls of the last three water-closets mentioned were made of lead, although there is no statement to that effect in the specifications.

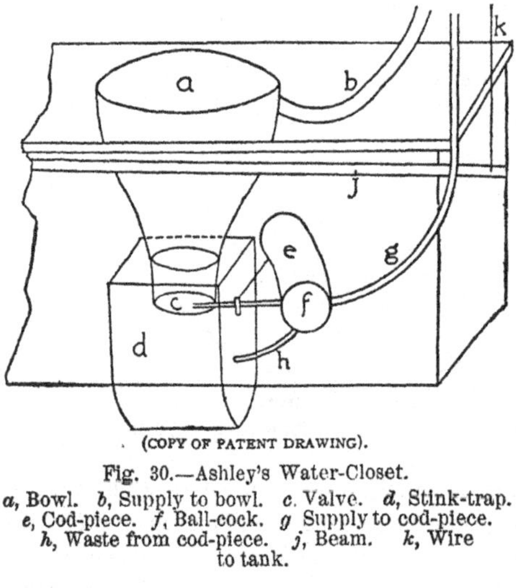

(COPY OF PATENT DRAWING).

Fig. 30.—Ashley's Water-Closet.
a, Bowl. *b*, Supply to bowl. *c*. Valve. *d*, Stink-trap. *e*, Cod-piece. *f*, Ball-cock. *g* Supply to cod-piece. *h*, Waste from cod-piece. *j*, Beam. *k*, Wire to tank.

Patents were issued to James Grease, 1782; David Riz, 1783; and Thomas Roundtree, 1789, for improvements on commodes.

In 1792 a patent was issued to John Ashley* for "a New Invented Self-Acting Water-Closet (Fig. 30), which constantly keeps itself sweet and clean and is upon a more simple construction than any heretofore discovered." This is a valve-closet supplied by a tank, the hopper being flushed by pressure on the seat. The specifications say: "Anyone sitting on the seat presses it down about an inch, which presses on the beam, and opens the sucker in the upper part of double cistern and lets the water in, and when the person rises off the seat the upper sucker closes and opens the lower sucker and lets the water down the two pipes; the one washes the bason [bowl] while the other fills the cod-piece and causes the air-ball to rise, the closer [valve] falls and lets out soil and water into said stink-trap, and by means of the small pipe the cod-piece is emptied of the water, the ball falls and raises the closer [valve], and prevents the effluvia ascending."

The common form of stink-trap is used under this closet, as we are told by the author, and in this way we gain as clear an insight into the form of trap used at that time as it is possible to do from such an imperfect drawing. It was probably something like our D-traps.

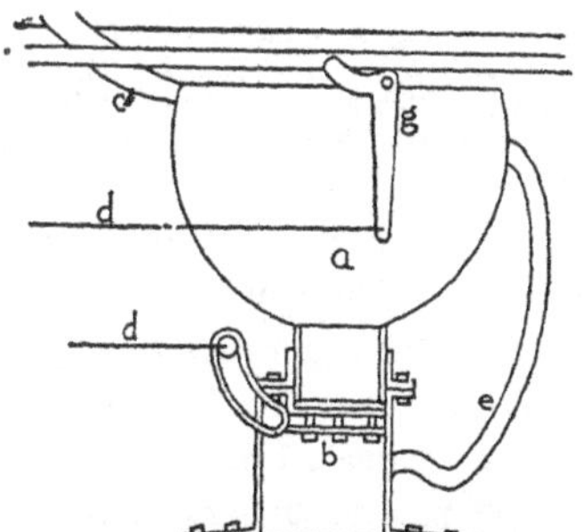

Fig. 31.—Binn's Closet.
a, Bowl. *b* Valve. *c*, Supply. *d* Wires to tank. *e*, Overflow. *f* Soil-pipe. *g*, Lever, worked by seat.

A patent† was issued to Thomas Binn, carpenter, 1793, for a valve water-closet (Fig. 31). It is very much like the Bramah closet, with the exception of the way the levers are arranged in connection with a complicated tank for the purpose of flushing the closet. Binn received a patent for an improvement on his closet in 1799. The improvement consisted in joining a bidet-pan to the bowl of the closet.

* *British Patent-Reports*, Vol, XXIV., No. 1875.

† *British Patent-Reports*, Vols. XXV. and XXXI., Nos. 1937 and 2352.

William Laws received a patent* in 1796 for an improved water-closet. He shows us a receiver and a top view of a valve which he calls a pan. He does not claim novelty for either the pan or the receiver, so it is possible that there may have been pan-closets in use before the Bramah closet was invented, and therefore it may be wrong to cast the odium of their parentage upon it. The

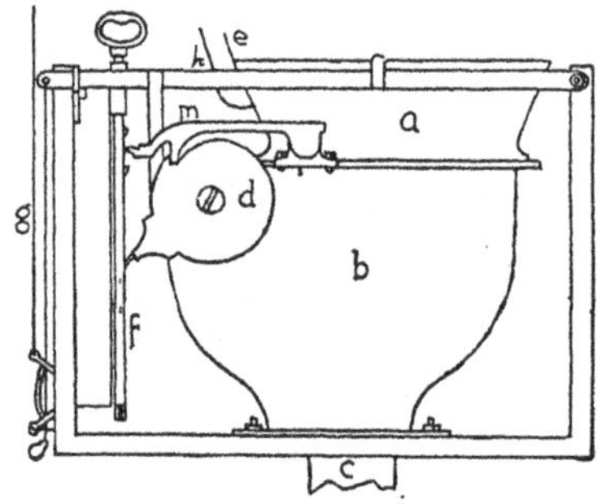

Fig. 32.—Law's Closet.
a, Bowl. *b*, Receiver or cesspool. *c*, Soil-pipe. *d*, Ratchet. *e*. Supply. *f*. Pull-rod. *g*, Wire to tank. *m*, Catch.

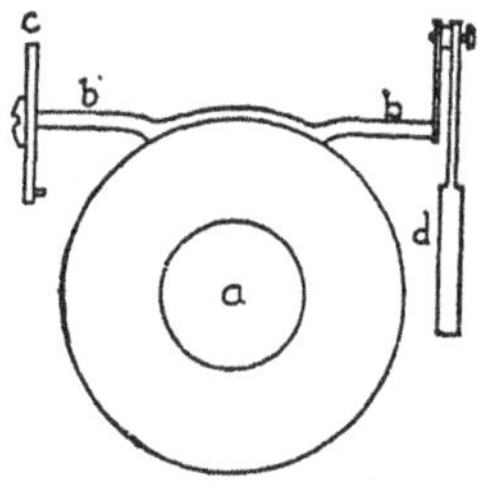

Fig. 33.—Details of Pan.
a, Pan-valve. *b*. Axes. *c*, Ratchet. *d*, Weight.

Laws closet is intended to act on the seat being relieved from pressure. The handle is patented in connection with a self-acting closet "*to put the self-acting part to rights, and may be found very useful at some time.*"....Another accommodation is a lever conveniently situated to act by the foot.....The weight on the axis of the pan turns back to a catch, that holds the pan fast to bear the water for the next action of the closet.

The receiver or cesspool is very much like those used on pan-closets at the present day.

French Patent Law began 1791. Between 1794 and 1800 (the date which I have assumed as the date for ancient closets to merge into modern closets) the French Government issued one patent for a water-closet to De Cœur, January 5, 1796.

There was no patent issued for water-closets in America until

* *British Patent-Reports*, Vol. XXVII., No. 2111.

1833 and 1835, and those were issues for portable water-closets or commodes.

Reviewing contemporaneous literature we find that water-closets before the nineteenth century were not considered a part of the plumbers' work. There is no mention of water-closets in "Art du Plombier," by C. M. de Gardette, an exhaustive work on the subject (1761–1779), while I find a description of the one described above in "L'Art Menuisier," by A. J. Roubo. S. S. Hellyer tells us that he could find no mention of them in the "Art of Building," published in London, 1734, or in Campbell's "Compendious View of all the Trades practiced in the cities of London and Westminster," published in 1747.

A point which strikes me as curious is that I do not find mentioned or shown among the earliest closets, the simplest form or hopper-closet. Was it because hopper-closets were not in use at this early date, or because they were considered too simple to put on record?

SECTION II.

MODERN CLOSETS.

THE subject of water-closets that were in use before the year 1800 was reviewed by the author in a paper read before the Sixteenth Annual Convention of the American Institute of Architects, printed in the preceding pages. In the following sections I will treat of the forms or patterns that have been used between 1800 and the present time, covering a period which has been remarkably prolific in mechanical inventions and contrivances.

It would be neither useful nor profitable to describe all the inventions for which patents have been issued on this device; but those will receive due attention that have either merit in themselves, or have merits or demerits that in the opinion of the author bear upon the many patterns of closets now in use.

Classification.—Water-closets may be divided into the classes in which they naturally fall. This mode of classification was first used by Mr. T. M. Clark, in his articles on "Modern Plumbing," published several years ago. The "valve closets" include all that have a valve whose points of contact with the bowl or additions thereto form a water-tight joint. The valve keeps a certain quantity of water in the bowl.

"Pan-closets" include all that have a dish-shaped basin or pan at the bottom of the bowl. The pan forms a water-seal with the bowl or a projection therefrom.

"Plunger-closets" include all that have a plug or plunger fitting over or into the entrance of the trap or soil-pipe; by this means any required amount of water is kept in the bowl.

Among "hopper-closets" I include all that have a simple bowl and no mechanical contrivances other than a water-seal to keep the sewer-air of the soil, drain or sewer pipes from entering the house.

Under the head of "latrines" are classed all rows of closet-bowls which are in reality one receptacle, having one outlet or junction with the soil-pipe, and a trap, plunger or valve common to them all.

There are a small number of water-closets that will deserve notice, which cannot be properly placed under either of the above classes.

Among the early inventions, by far the greater number were for valve-closets. It is probable that hopper and pan-closets were in use before the invention of valve-closets, while there is positive evidence of the plunger-closet being in existence and use before the invention of either the Cummings or Bramah closets.

VALVE-CLOSETS.

Valve-closets may be treated of under the head of sliding and hinged valves, the latter being by far the most useful and numerous type of this class.

It is sometimes difficult to decide, when the pan is intended to fit tightly against the receiver, whether a closet belongs properly to the valve or the pan class; so under this head I describe only those that have valves properly so called. The object of the valve is to intercept any sewer-air or organic germs that may come from the traps, soil, drain or sewer-pipes, and at the same time to keep the bowl filled with a certain amount of water, about half-filled in most cases. The large volume of water deodorizes the fæcal matter discharged into it, and at the same time it prevents the bowl from becoming soiled.

In connection with these closets we always find a compartment between the bowl and soil-pipe, in which the valve either slides or works on its hinge or spindle. This closet must of necessity have an overflow.

In early times these closets were supplied from a special cistern placed over and above the closet; now they are supplied either from the main supply or from special cisterns. The water-supply is generally turned on into the bowl, and at the same time the valve of the closet is opened by one and the same lever. The largest number of patents are issued for novelty in the manner of combining these cranks, wires and levers with the hand-pull; in some cases the inventions are only remarkable for the complication of their mechanism, the inventors forgetting that the foundation of their usefulness is simplicity.

Sliding-Valves.—The sliding-valve closets have the first closet patented in Great Britain in their class. (See Cummings' closet).

Lucknow's Closet.—The only patent issued for a valve similar to the Cummings' valve was issued to John Lucknow in the year 1854 by Great Britain. This closet has a sliding-valve, with a circular hole of the same diameter as the soil-pipe in one end. When the valve has been drawn back as far as possible, the two holes coincide, and the waste matter in the bowl is precipitated into the soil-pipe The sliding bar which operates the valve is connected with the hand-pull.

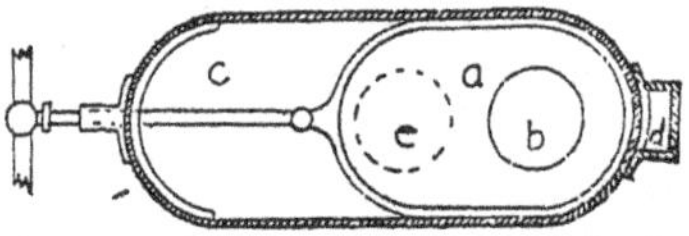

Fig. 34.—Valve and Receiver of Lucknow's Closet.
a, Slide-valve. *b*, Hole in valve. *c* Receiver. *d* Overflow. *e* Soil-pipe. *f*, Pull-rod connected with hand-pull.

Viney's Closet.—In the year 1824 there was a valve-closet invented in England by James Viney, that was the first of another type belonging to this class. The valve in this closet might be called a sliding gate, that slides up and down in a vertical position. The valve has grooves or guides in which it works. The bowl is kept partially full of water when the valve is closed. The compartment in which the valve works is ventilated, and the branch from the water-closet has a flap-valve where it enters the soil-pipe. The vent and overflow pipes run directly through the wall and have their ends open on the outside of the building. This closet has a minimum of space devoted to the valve-compartment, having a large body of water to

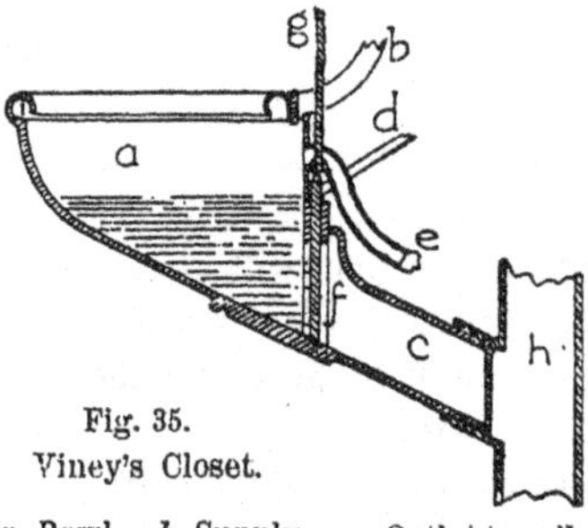

Fig. 35.
Viney's Closet.
a Bowl. *b*, Supply. *c*, Outlet to soil-pipe. *d*, Vent to valve-compartment. *e*, Overflow. *f* Valve. *g*. Rope or chain. *h*, Soil-pipe. *i* Flushing rim.

flush the drains and no water-seal trap under it, so it would be an excellent closet according to Mr. Norman Shaw's ideas. A valve of this kind and apparent weight would be liable to fail from waste matter catching in the grooves, and in this manner preventing the gate taking a proper seat. In this closet I note, also, the earliest flushing rim that extends around the bowl in an annular ring, and has an outlet so arranged as to flush and scour all parts of the bowl.

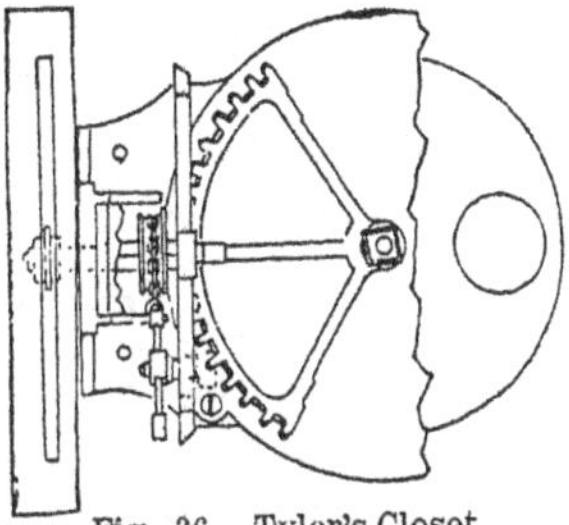

Fig. 36.—Tyler's Closet. Top view with bowl and part of valve-chamber removed.

Hanson's Closet.—A patent was issued by the United States in 1882, to E. Hanson, for a sliding valve that is intended to slide across the outlet of the closet, retaining its vertical position, similar to the valve on the closet of S. N. Grubb, but sliding up and down vertically, instead of horizontally, as in the case of the latter. (See Fig. 40).

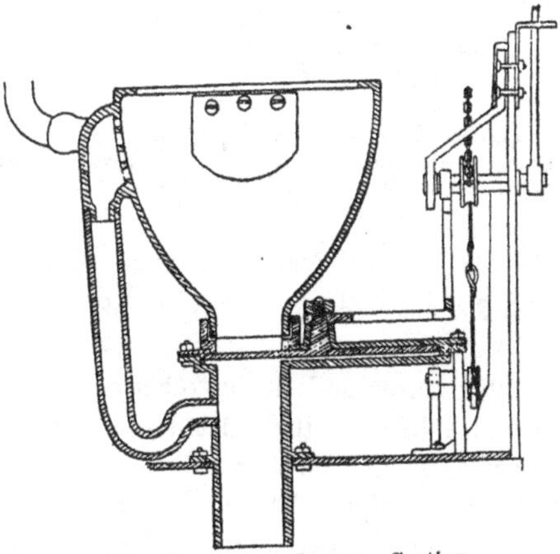

Fig. 37. Tyler's Closet.—Section.

Tyler's Closet.—In 1829, Hayward Tyler, a Quaker, and a brass-founder of London, invented a closet with a horizontal sliding-valve that rotates on a pivot. The valve consisted of a circular disc with a hole in one side of it, which had the same diameter as the soil-pipe. When the disc is turned so that the hole in it coincides with the opening in the soil-pipe or trap, the contents of the bowl are emptied directly into the soil-pipe. Motion is imparted to this valve by a hand-lever connected to a vertical toothed quadrant, that has its axis at right angles to, and its teeth geared with, a horizontal quadrant. The axis of the horizontal quadrant is also the pivot on which the valve turns, so that any movement of the hand-lever will cause the valve to rotate on its centre. In this closet the valve-compartment is small and it does not act as a receiver for filth, the waste matter passing directly into the soil-pipe or trap; but the machinery is very complicated, and for this reason liable to get out of order and to require intelligent supervision, which is rarely expected from the householder, and never from the servants. I note three water-closets invented in the United States with pivoted horizontal sliding-valves: S. N. Grubb, 1879; F. Watson, 1880; J. Robertson, 1881.

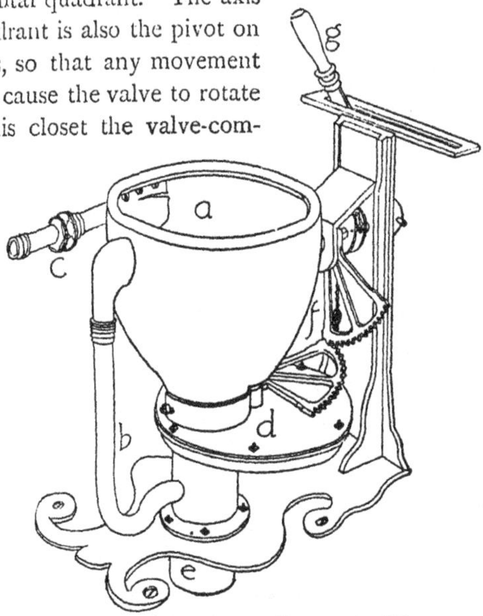

Fig. 38.—Tyler's Closet.—Perspective View.
a, Bowl. *b*, Overflow. *c*, Supply. *d* Valve-compartment. *e*, Entrance to soil-pipe in trap. *f*, Geared quadrants. *g*, Lever.

Grubb's Closet.—In Grubb's closet and in the other two mentioned above, the only difference being in their details and in the manner of working the valves, is a disc that has a diameter a little greater than the opening in the soil-pipe. This disc is attached to one end of an arm, while at the other end is the pivot on which the valve rotates. The valve has springs to press it tightly against its seat, and a branch from the supply-pipe to wash off the valve and to wash out the valve-compartment. The supply-valve is worked by the same pull that rotates or slides the valve.

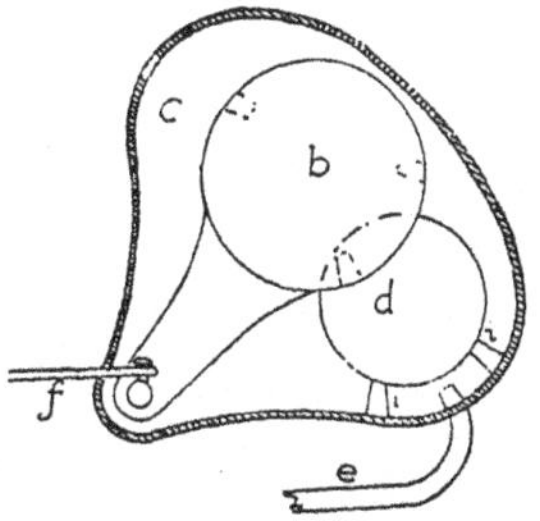

Fig. 39.—Section of Valve-chamber.

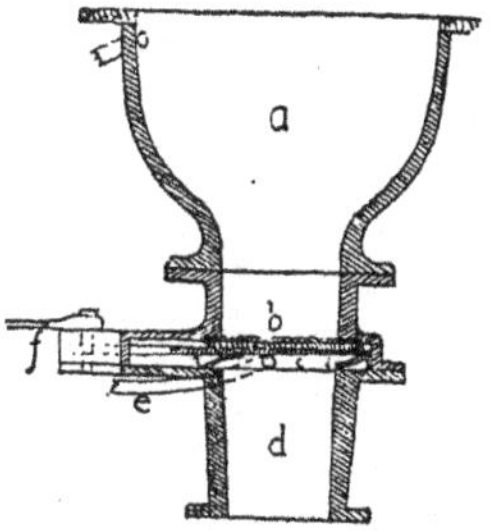

Fig. 40.—Section.

Grubb's Closet.

a, Bowl. *b*, Valve. *c*, Valve-compartment. *d*, Entrance to soil-pipe or trap. *e* Supply-pipe to valve-chamber. *f*, Rod connecting with hand-pull. *i*, Springs.

Wilkin's Closets.—There was a curious sliding or turning valve closet invented in England in 1846 by one J. W. Wilkins. The receiver was a cylindrical box in which worked a cylindrical valve,

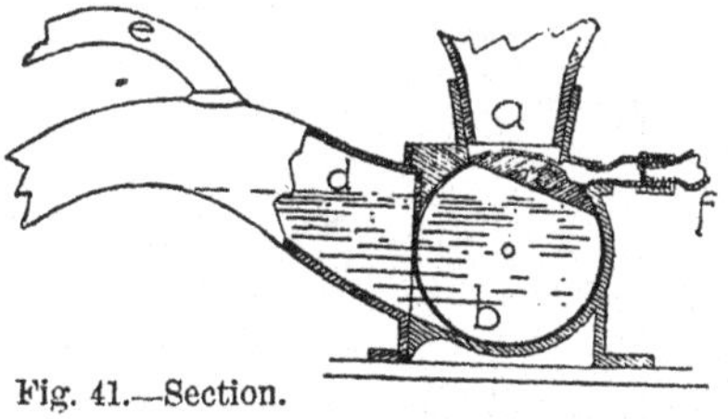

Fig. 41.—Section.

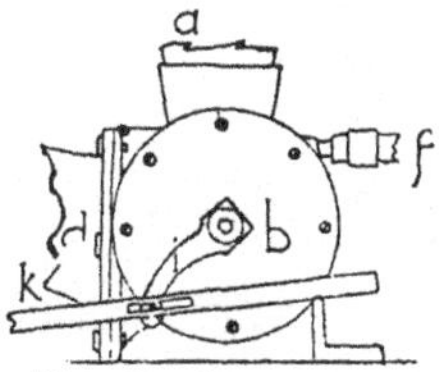

Fig. 42.—Side View.

Wilkin's Closet.

a, Bowl. *b*, Receiver. *d*, Trap. *e*, Vent-pipe. *f*, Supply to valve. *i*, Crank. *k*, Lever connected with pull-rod.

a section only being solid. This solid part of the valve was kept in position by a weighted lever, so as to form a bottom for the the bowl of the closet; when the lever was raised, it turned the solid part of the valve away from the bottom of the bowl, dropping its contents into the receiver. In connection with this closet I find for the first time a vent-pipe connected with the crown of the trap. The specifications claim it as a novelty.

Norton's Closet.—The two following closets, which work on the same principle as the Wilkins, were invented in the United States in 1876 and 1882 respectively. Norton in his invention applied the principle of a ground-cock to the valve of a water-closet. The valve is simply a piece of conical metal with a hole the same diameter as the soil-pipe, which, when turned so that they will be concentric, will allow the waste matter in the bowl to pass into the soil-pipe. The valve is kept in position by the aid of a weight and spring, as is shown in the cut.

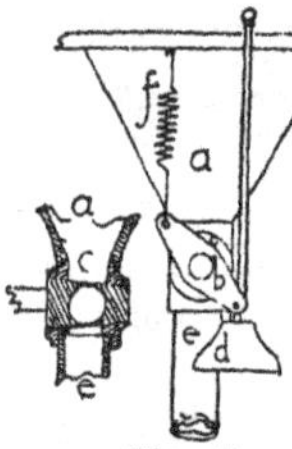

Fig. 43.
Norton's Closet.
a, Bowl. *b*, Crank.
c, Valve. *d*, Weight.
e, Soil-pipe. *f*, Spring.

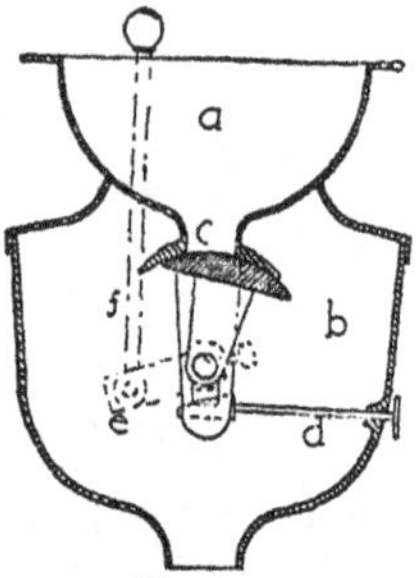

Fig. 44.
Daggett's Closet.
a, Bowl. *b*, Receiver.
c, Valve. *d*, Set-screw.
e, Crank. *f*, Pull-rod.

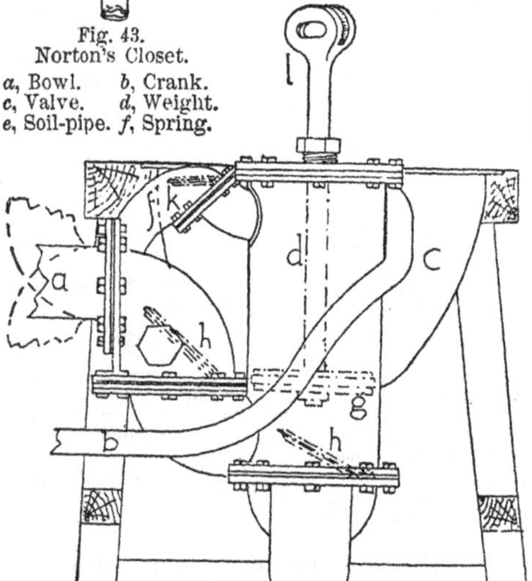

Fig. 45.—End View.—Downton's Closet.
a, Discharge-pipe. *b*, Supply-pipe. *c*, Bowl. *d*, Cylinder.
f, Vent-pipe. *g*, Piston. *l*. Place for handle.
h, Upward valve. *i*, Hand-hole. *k*, Vent.pipe.

Daggett's Closet.—Daggett's closet in the action of the valve is almost identical with the one invented by Wilkin's. The valve is a segment of a cylinder that revolves on its axis. This valve is operated by means of a crank attached to the hand-pull. When the valve has revolved to a certain distance, it leaves the orifice at the bottom of the bowl open. This closet has a screw with the handle on the outside of the receiver, so the valve may be made to press more firmly against its seat if it should become loose.

Sliding-valve closets seem to have been in actual use to a very limited extent, the reason probably being that the inventors or manufacturers found that they would not work in practice, it not being practicable to make the valve slide when and where the operator desired. It would be almost impossible to adjust the parts so they would have a water-tight joint for any length of time. The hinged valve-closets are generally simpler in their construction; a tight joint with this form of valve is more practicable, and there are a large number of the hinged type of valve-closets in use at the present day in all parts of the world.

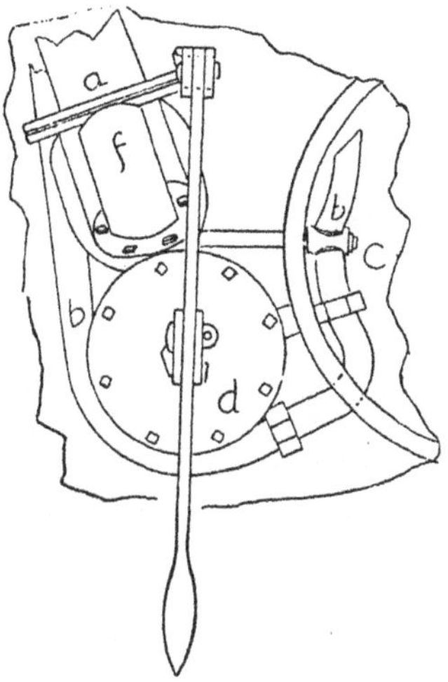

Fig. 46.—Partial View of Top. Downton's Closet.

HINGED-VALVE CLOSETS.

Among closets that have hinged valves, or valves working on a spindle, I find one type in which the valve opens upward, another outward, and another downward in a direction relative to the bowl of the closet.

Downton's Closet.—Taking under consideration first the closets in which the valves open upward or toward the bowl, I find the first invention of this type was made in Great Britain in 1825, by J. Downton, for what he calls a "pump-closet." In this closet the matter in the bowl is drawn by the upward action of the piston into a cylinder, where it is retained by a valve that opens only in an upward direction. When the piston is pressed down by a handle, the fæcal and other matter is discharged in any direction that may be desired. The return to the cylinder of the matter discharged is prevented by a flap-valve opening only in an upward direction. To enable the piston to work easily, the cylinder has an air-inlet near the top. The supply-valve is connected with the handle that works the piston. Messrs. Tylor & Son, of Newgate Street, London, manufacture Downton's closet at the present day, with

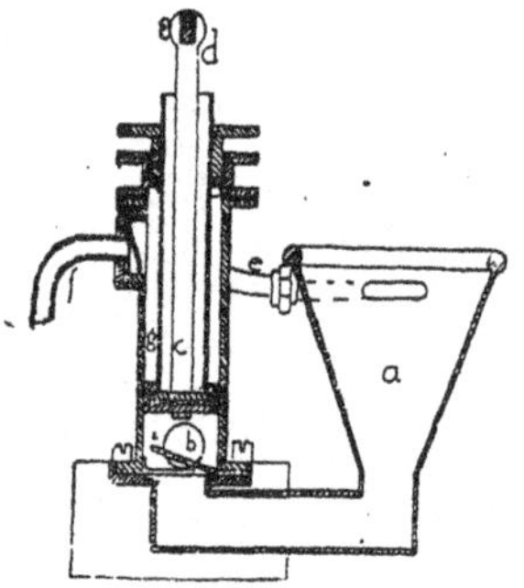

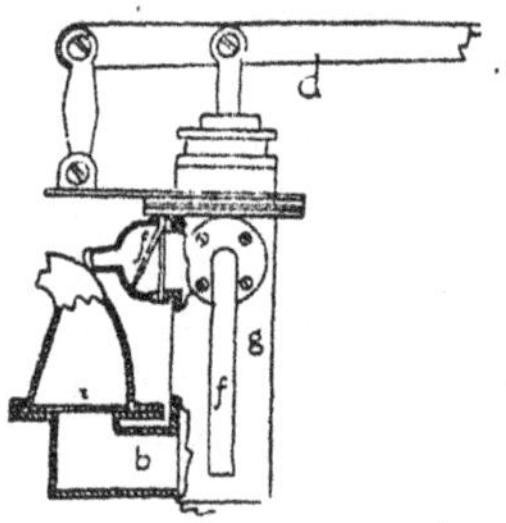

Fig. 47.—Section. Sands's Closet. Fig. 48.
a, Bowl. *b*, Discharge-pipe. *c*, Piston. *d* Lever. *e*, Supply. *f* Air-pipe. *g*, Cylinder. *i*, Valves.

what they consider an improved form of lever-handle, "which works the double-action pump and water-tap at one action....It is equally effective above or below the water-line."

Sands's Closet.—In the United States, in 1874, one Sands invented a closet of this tyye that is very similar to Downton's. This device is also intended to be placed below the water-line or the point of discharge. By raising a lever, the excrementitious matter in the bowl is drawn by the suction of a piston directly into a cylinder. When the lever is pressed down, the waste matter may be discharged in any direction required, through the soil-pipe previously placed in the proper position. The flap-valves, of which there are two, opening in an upward direction, prevent the return of the waste matter to either the bowl or the cistern.

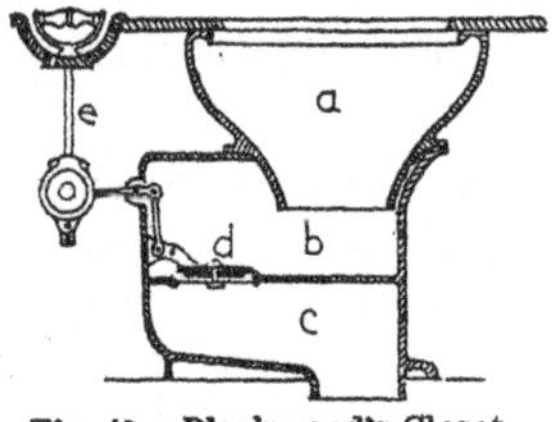

Fig. 49.—Blackwood's Closet.
a. Bowl. *b*. First receiver. *c*, Second receiver. *d*, Valve. *e*, Hand-pull.

Blackwood's Closet.—I note, as the only instance of this type of closet being intended for general use, one invented in this country by W. Blackwood in 1881. In this closet the matter which drops into the bowl would go directly into a large compartment or receiver. The valve is opened upwards by means of a combination of levers connected with the hand-pull. When the valve is raised, the water and excreta would find their way into a second receiver. The first compartment, which opens directly into the room through the bowl, would without doubt get and remain in a very filthy condition.

Bunnett's Closet.—The first closet invented with a valve which opens in a direction outward from the bowl was designed by an English engineer named Joseph Bunnett, in 1846. This closet was designed to be placed below the water-line or point of discharge.

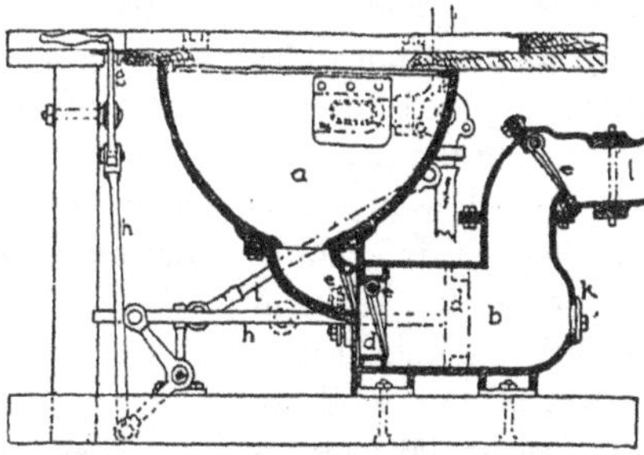

Fig. 50.—Section.—Bunnett's Closet.
a, Bowl. *b*, Piston chamber. *d*. Piston. *e*, Valves. *f*, Supply-pipe. *g*, Hand-lever. *h*. Piston-rod. *i* Rod for operating supply-valve. *k*. Inspection-hole. *l*, Soil-pipe. *m*, Fan. *n*, Connecting-rod.

This closet appears to be simpler in its construction than the Downton closet, and for that reason would be better for the purpose which that is intended to fulfil. The outlet at the bottom of the bowl and the piston which works in the receiver both have flap-valves, which open from the bowl in an

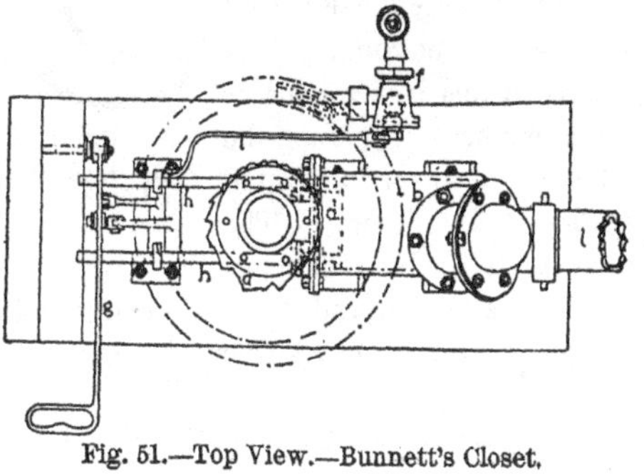

Fig. 51.—Top View.—Bunnett's Closet.

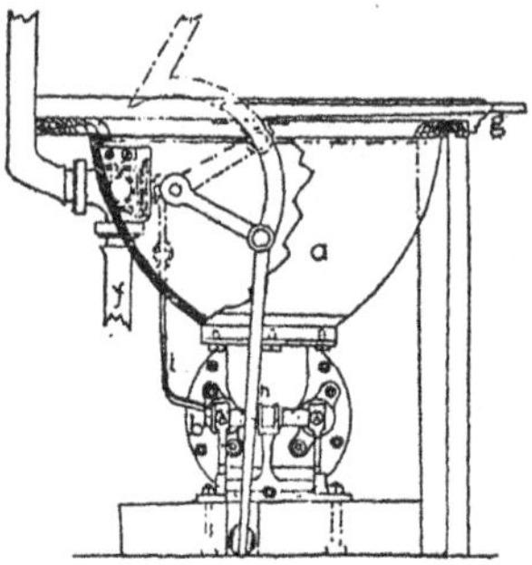

Fig. 52.
Side View.—Bunnett's Closet.

outward direction. The waste matter from the bowl will descend into the receiver by the action of gravity; thence it would be forced by the action of the piston through a third valve into the soil-pipe. The last-mentioned valve is for the purpose of preventing the return of the discharged matter into the receiver. The outlet of the soil-pipe may be carried in any direction desired. In this manner the discharge of the excreta may be governed to suit circumstances. Water-closets of this type may in rare instances become necessary; for instance, in cases where it is required to put a water-closet below the point of connection with the sewer.

When used it would be advisable to run the soil-pipe higher than its point of juncture with the sewer, so there could be no back-flow from the sewer in case the valves should leak—and valves of this kind are liable to leak at the most inopportune time.

Armstrong's Closet.—In 1848 a closet of this type and class was patented in England by John Armstrong. The novelty in this closet consists in the manner of opening the valve, by means of a slotted quadrant and cam. The valve is lifted by a cam attached to a vertical rod, the cam being bolted to the rod so that the valve will be lifted when the cam is turned on its greatest axis.

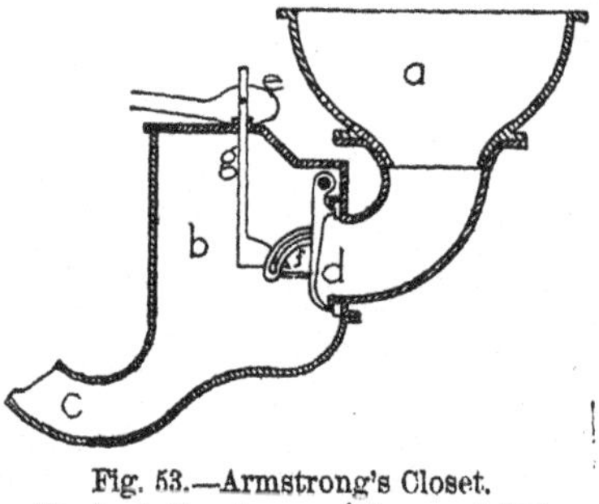

Fig. 53.—Armstrong's Closet.
a, Bowl. *b*, Receiver. *c*, Trap. *d*, Valve. *e*, Cam. *f*, Slotted quadrant. *g*, Connecting-rod.

Feilding's Closet.—Another English patent was taken out for a closet, in which the valve opens outwardly, in the year 1855, by Fletcher Feilding.

There is a reservoir below the bowl that forms a water-seal trap for the overflow. This reservoir would become very filthy from deposits and stagnant water, as there is no means of flushing it. The valve has a leather washer, and is lifted by means of a crank that works in a U-shaped attachment to the valve.

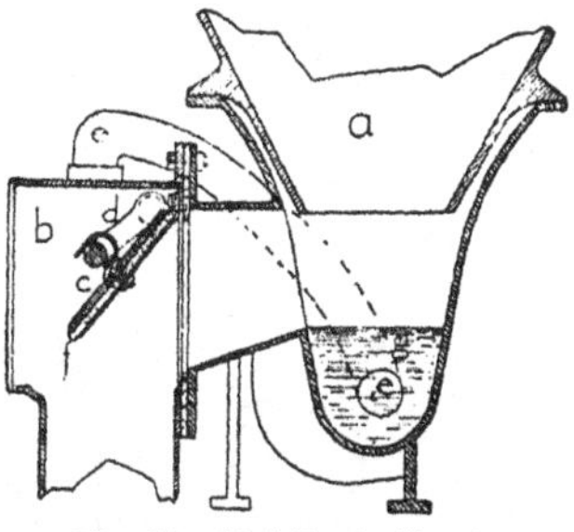

Fig. 54.—Feilding's Closet.
a, Bowl. *b*. Receiver. *c*, Valve. *d*, Crank. *e*, Overflow. *f*, Washer. *g*, Trap to overflow.

Bean's Closet.—Bean's closet, used in Scotland, may properly be classified as one of this type of valve-closets. The valve seems intended only for the purpose of retaining a large amount of water in the bowl, so that by opening the valve quickly the whole amount would be discharged suddenly into the soil-pipe, for the purpose of scouring it more thoroughly. There is an opening above the valve which is an overflow. This closet is simply a short hopper-closet with its trap, the valve being placed where the trap discharges into the soil-pipe.

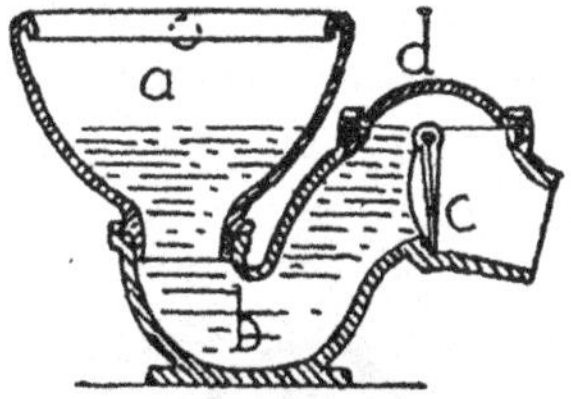

Fig. 55.—Bean's Closet.
a, Bowl. *b*, Trap. *c*, Valve. *d*, Inspection-cover.

Dummis's Closet.—F. Liger, in the French work on water-closets, urinals, etc., published in 1875, describes, among other closets, a valve-closet of this type, which he tells us was at that time (1875) in common use in France. This closet has a chamber or compartment at one side, very similar in appearance to the ones used in connection with plunger-closets. In this chamber there is a weight attached to the rod of the hand-pull and connected with the valve by a short hinged arm. The weight, which moves only in a vertical direction, keeps the valve pressed tightly against its seat, and unless the weight is lifted, it would be impossible for the valve to open. F. Liger says of this closet, that it is one of the best, uniting all the conditions required for health, and that it is odorless. I can see no reason why it is exempt from the faults of the class to which it belongs.

Dummis invented, in connection with his closet, a device for separating the liquid from the solid excreta. At the point where the trap enters the soil-pipe it is divided into two branches. A grating or perforated plate is placed over the outlet nearest the closet, while a valve opening outward closes the one furthest from the closet bowl. This valve is held in position by two weighted levers. When the waste matter is discharged from the bowl, it passes through the trap, but is checked by the valve and stopped directly over the grating. The liquid matter would then pass

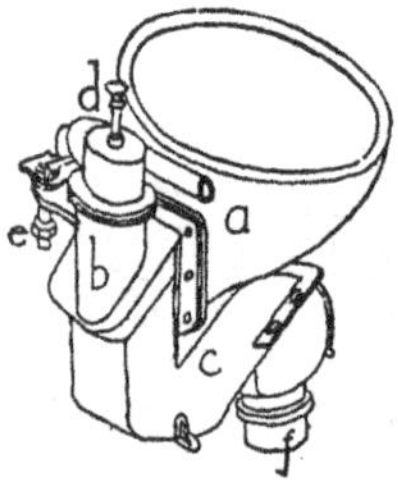

Fig. 56.

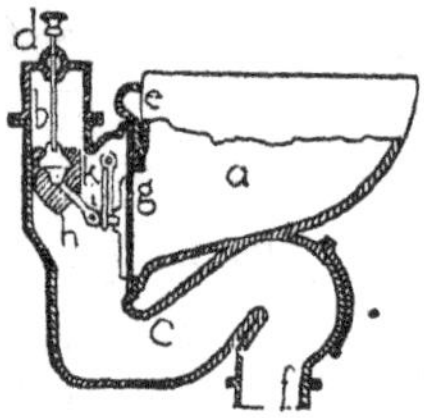

Fig. 57.

Dummis's Closet.

a, Bowl. *b*, Valve-compartment. *c*, Trap. *d*, Hand-pull. *e*, Supply-pipe. *f*, Soil-pipe. *g*, Valve. *h*, Weight. *i*, Connecting-arm. *k*, Hinge,

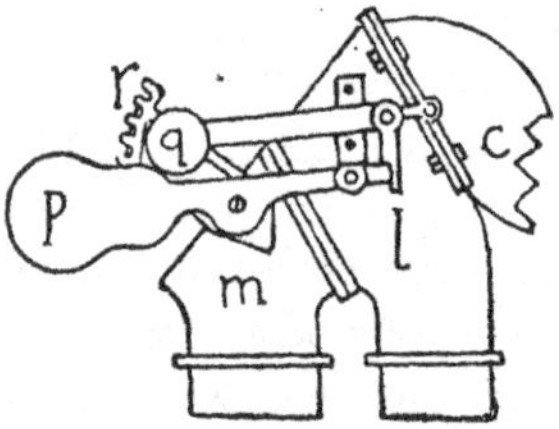

Fig. 58.—Side View.—Dummis's Closet.

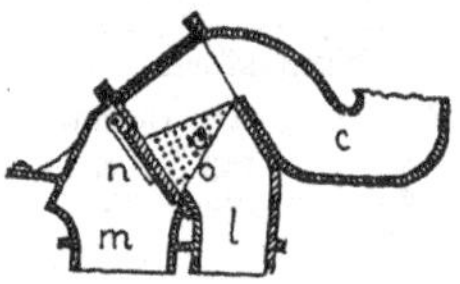

Fig. 59.—Section.—Dummis's Closet.

Device for dividing liquid from solid excretra.

c, Water-closet trap. *l*, Soil-pipe for liquid matter. *m*, Soil-pipe for solid matter. *n*, Valve. *o*, Perforated plate. *p*, *q*, *r*, Combination of weights and levers for valve.

through the perforated plate into the branch of the soil-pipe nearest the closet bowl, leaving the solid matter to be discharged into the other branch of the soil-pipe by opening the valve. This would be accomplished by raising the weighted levers on the outside. By thus dividing the liquid from the fæcal matter, its manurial qualities can be more easily utilized, and I find that the French inventors almost invariably have this point in view in making their inventions. This practice must often be a detriment to comfort by reason of the unpleasant odors generated, and to cleanliness and health, for as soon as the liquid matter is separated from solid matter, all the scouring and deodorizing qualities of water are lost. The soil-pipes must become very foul, as the solid excreta passes, in many instances, through several stories before reaching its final receptacle in a large barrel or iron can (*Fosse mobile*).

Carr's Side-Outlet Valve-Closet.—W. S. Carr received letters patent in this country for a closet similar to the French closet in the arrangement of the valve and the weight which holds it in position.

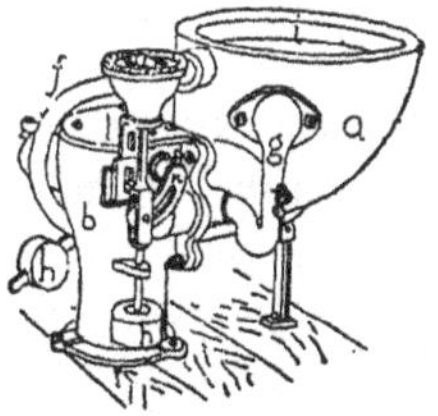

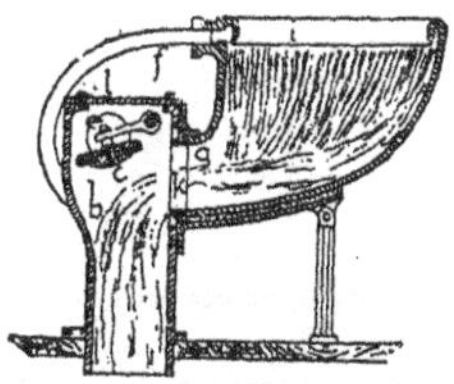

Fig. 60.—Perspective. Doulton's Trapless Closet. Fig. 61.—Section.
a, Bowl. *b*, Valve-compartment. *c*, Valve. *e*, Vent. *f*, Supply. *g*, Overflow. *h*, Weights. *i*, Flushing-rim. *k*, Valve-seat. *l*, Removable cover. *n*, Valve-crank.

Doulton's Trapless Closet.—Probably the best of this type, and I think the best valve-closet, provided it has a siphon-trap below it, properly vented, is the "trapless" closet manufactured by Doulton & Co., Lambeth, London. This closet has a flushing-rim that gives an equal distribution of water to all parts of the bowl, by means of small holes in the rim which encircles the top of the bowl. The manufacturer says: "The construction of the closet renders the usual complication of levers, both for the supply and discharge valves, unnecessary, as both are worked from the same spindle with direct action at each end, a very small weight being necessary to render the discharge-valve tight." There is a metal rim for the valve to fit against, while the valve has a yielding substance such as rubber where it comes in contact with the seat, and an earthen face.

The overflow from the bowl to the valve-compartment is sealed by a U-trap. The bowl is furnished in different ornamental patterns and attached to an iron receiver or valve-chamber by means of bolts. The perspective view shows the simplicity of the mechanism. The hand-pull is connected to the spindle, which turns the valve by a slotted crank. The weight which holds the valve in position is attached to the lower end of the pull-rod. The valve-chamber of this closet has a large vent-pipe, and a cover held on by screws that may be easily removed for the purpose of cleaning out this compartment. It is important that these inspection-covers should fit air-tight; when once removed, they are generally put back in a careless manner, and in the case of a trapless closet this would allow free communication between the soil-pipe and the house. Doulton's closet can receive its water directly from the water-main, through his patent supply-valve; or it can be flushed from a cistern, through a pipe with an inch and a quarter diameter.

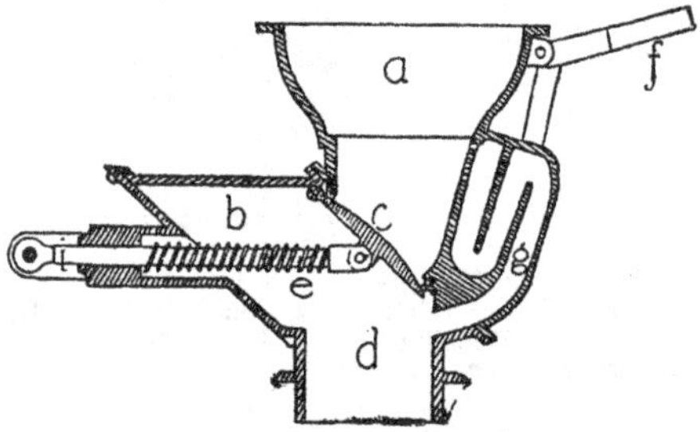

Fig. 62.—Pohley's Closet.
a, Bowl. *b*, Valve-chamber. *c*, Valve. *d*, Soil-pipe. *e*, Spring. *f*, Lever. *g*, Overflow. *i*, Rod.

Pohley's Closet.—In this country, in 1877, a valve-closet of this type was invented by F. Pohley. The valve is held in its place by a spring that encircles the rod by which the valve is drawn back or held in position. The valve, of metal, fits against a pliable band which is let into the bottom of the bowl. The valve is opened by an L-shaped lever, and the bowl has a simple siphon overflow.

Tylor & Son's Valve-Closet.—Tylor & Son, of Newgate Street, London, invented, in 1878, a closet in which the valve opens in the same direction as the closets described above. The outlet to the bowl is at the bottom; a projection from this opening turns at right angles or forms a quarter-bend, and on the end of this projection the valve finds its seat. The overflow enters the valve-chamber immediately back of the valve, and is sealed by dipping into a box formed in the receiver. The valve with a "rubber or other suitable material, washer closes against the brass or other metal valve-seat." This valve-seat is clamped to, and made to project beyond the bowl, as shown in the detail cut.

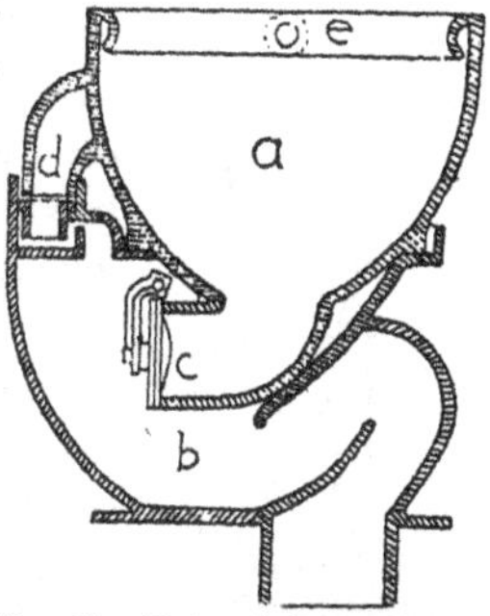

Fig. 63.—Tylor & Son's Side-outlet Valve-Closet.
a, Bowl. *b*, Trap and receiver. *c*, Valve. *d*, Overflow.

The type of valve-closets which I have last described has an advantage over other closets of this class, as well as over plunger-closets, in the position of the valve and its chamber. The waste matter passes directly into the soil-pipe without passing through the valve-chamber proper, as it does in other closets of this class, and the water does not stand in this chamber, coating it with foul deposits, as is the case in plunger-closets. With this form of closet there is very little opportunity for the back of the valve to become foul, as neither the water nor waste matter ever reaches it.

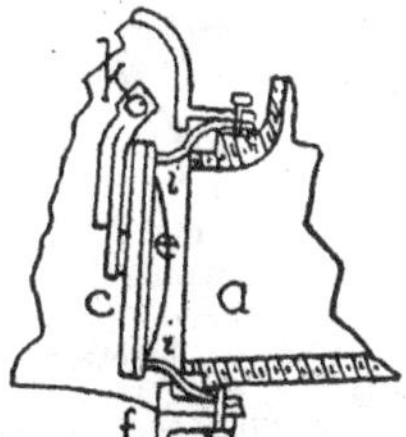

Fig. 64.—Tylor & Son's Side Outlet Valve-Closet.—Detail of Valve.
a, Bowl. *c*, Valve.
e, Earthenware face.
f, Yielding washer.
i, Metal seat.
k, Spindle.

At the present day Doulton & Co.'s "trapless" closet and Bean's closet are used in Great Britain, while Dummis's is used in France. There are no closets of this type manufactured in this country, and none in use, unless a small number of the English closets have been imported for parties who have been pleased with their construction.

VALVES THAT OPEN DOWNWARD.

I now come to the last type of valve-closets, and this form of the class has been in use since the beginning of the eighteenth century, with slight variations in the arrangement of their mechanism, and in the materials used in their construction. These closets, in which the valve opens in a downward direction from the bowl, have their prototype in the Bramah closet invented more than a hundred years ago. English manufacturing firms, notably Tylor & Son, Underhay, and Jennings, manufacture closets under that name at the present day. Doulton & Co. manufacture a valve very much like the Bramah, adding a weight to the end of the lever, which moves the valve, and a vent-pipe to the receiver.

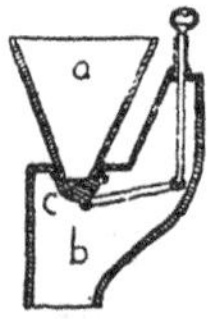

Fig. 65.
Early French Valve-Closets.
a, Bowl. b. Receiver. c, Valve.

Early French Valve-Closet.—Liger gives a description of a closet of this type which was first brought into use in France about the year 1823. This closet had a metal valve which was intended to fit tightly against the bowl. The valve was connected with the hand-pull in the simplest manner, by means of a short hinged arm. Judging from appearances there was no weight to keep the valve in position. The pull-rod and connecting arm were both within the receiver. The opening where the rod passed through appears to have received special attention.

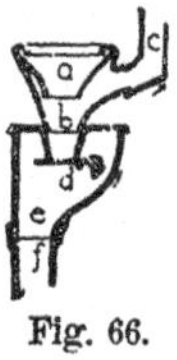

Fig. 66.
Flament's Closet.
a, Bowl.
b, 1st Receiver.
c, Vent.
d Balance-valve.
e, 2nd Receiver.
f, Soil-pipe.

Flament's Closet.—Another closet used in France was invented by M. Flament. The bowl, with a flushing-rim, is set directly into a hopper or receiver. This hopper has a large vent, which is intended to run into a warm flue, if one is convenient; otherwise, into the outer air with a ventilating cowl on top, M. Flament designing a cowl in connection with his closet. At the bottom of this receiver, having a seat on a projection therefrom, is a balanced valve that works in a second receiver or valve-chamber. The valve is hinged, and on the side of the hinge opposite to, and forming a part of the valve, is a weighted projection that would cause the valve to rest firmly against its seat. When the first receiver has become filled or partially filled with water or excrementitious matter, the balance of the valve would be overcome and the waste matter dropped into the second receiver and soil-pipe. In the cases mentioned above, the valve is intended to keep disagreeable odors from entering the house. They are not intended to be placed over a siphon-trap.

The following closets are almost without a single exception in common use at the present day in different parts of the world.

Carr's Valve-Closet.—W. S. Carr, of New York, invented a simple valve-closet of this type in 1868. The bowl is set in a small

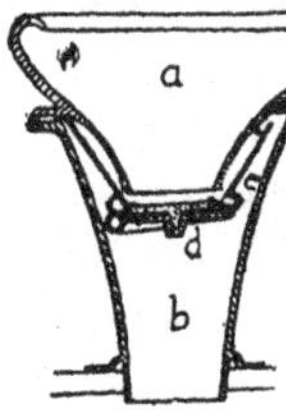

Fig. 67.

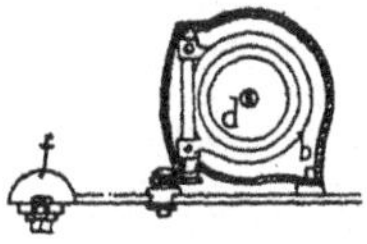

Fig. 68.

W. S. Carr's Closet.

a, Bowl. *b*, Receiver. *d*, Valve. *e*, Overflow. *f*, Lever.

hopper and the bottom of this hopper forms the valve-seat. The overflow is in the space between the hopper and bowl. The valve is held in position by a weighted lever. The inventor made some improvements on this closet in 1875, calling the improved closet the "American Defiance Closet." The novelty consisted in having the bowl and overflow made in one piece of earthenware, which is bolted to the receiver. The receiver is enamelled, and the part of the valve that shows in the bowl is also made of earthenware, closing against a rubber packing. Henry Huber & Co. now manufacture this closet with an opening at the top of the overflow for a vent-pipe, and they also furnish the bowls with a vent pipe from the closet bowl.

Peters's Closet.—Messrs. Peters & Donalds, of Glasgow, Scotland, manufacture a valve-closet in which the valve opens downward. The bowl of this closet is set in a small hopper which is placed above and forms a part of the receiver. The trap or water-seal to the overflow in this case is double, and has a vent-pipe from the crown of the trap, which would prevent the trap from being siphoned by the discharge from the bowl.

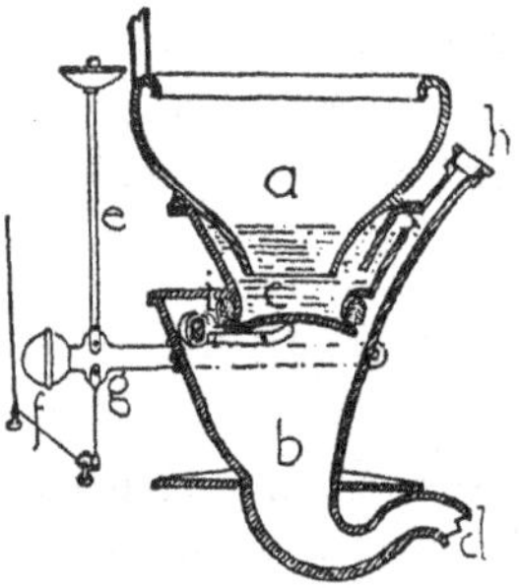

Fig. 69.—Peters's Closet.

a, Bowl. *b*, Receiver. *c*, Valve. *e*, Hand-pull. *f*, Weighted lever. *g*, Wires to tank. *h*, Vent to overflow. *i*, Rubber-ring (valve-seat). *k*, Trapped overflow.

J. Bailey Denton considers this one of the best closets in use, and in describing it says: "The valve arrangement of this closet is composed of a brass disc which is closed against an india-rubber ring by means of a projecting arm cast on a spindle. By this arrangement, the patentee states, the disc is closed perfectly tight against any uneven surface which at any time may present itself." The valve is held in position by the simplest form of weighted lever, while the supply-tank is connected with the closet by means of wires and bell-cranks, motion being imparted to both at the same time by the hand-pull. The trap, which is under the floor and formed in one piece with the receiver, is not vented. Letters patent were issued for the above closet to Peters & Peters, in 1871.

Hellyer's Valve-Closet.—S. S. Hellyer, of London, received patents, in 1875, for a valve-closet in which the valve opened in a downward direction from the bowl. The bowl is attached to the receiver by set-screws, and the receiver is enamelled. The part of the valve which shows from the bowl is earthenware, under which, and with a larger diameter, is a disc of rubber, leather, or other pliable material, which has a seat against a metal rim, in this manner forming a water-tight joint. The rubber disc rests on a third

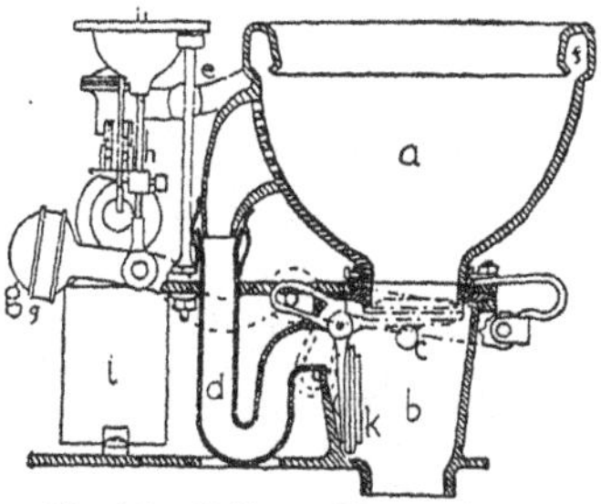

Fig. 70.—Hellyer's Valve-Closet.
a, Bowl. *b*, Receiver. *c*, Vent. *d*, Overflow. *e*, Supply. *g*, Weighted lever *h*, Hand-pull and stud for opening supply-valve. *i*, waste preventer. *k*, Valve.

disc of brass or some other suitable metallic substance. The valve is held firmly against its seat by the usual weighted lever, but the end of this lever, instead of being bolted so as to work on a fixed axis, is hinged or bolted to a spring. By this means the valve would be less liable to injury from violent jerks at, or careless dropping of, the hand-pull. The overflow enters the receiver behind the valve, and is supposed to keep this part of the valve and receiver clean. The receiver is ventilated, enamelled, and small, being only large enough to permit the valve to work properly. The water is admitted to the bowl through a flushing-rim, and is taken from a special cistern, or directly from the water-main. In the illustration the supply-valve is connected to a waste-preventer, and is operated by a small adjustable stud attached to the hand-pull by a set-screw.

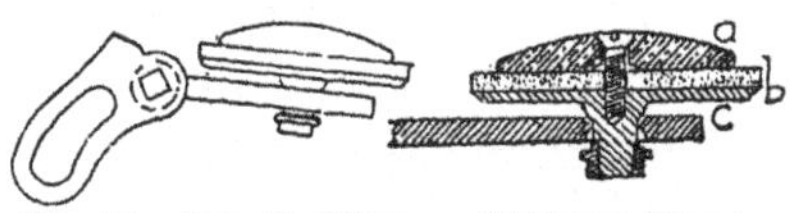

Fig. 71.—Detail of Valve.—Hellyer's Closet.
a, Earthen, or porcelain top.
b, Rubber, or metal disc.
c, Brass, or other metal back.

Underhay's Valve-Closet.—F. G. Underhay, of London, manufactures the Bramah, with slight changes. He also manufactures another simple valve-closet, in describing which he writes: "The bottom valve shuts against a ring of thick India-rubber bedded into a metal seating, effectually preventing leakage from unsoundness of valve—the great drawback to the old valve or Bramah closet. A good flush of water is obtained, no matter how carelessly the handle is pulled up or suddenly let down.... As the connection is already made between the supply-valve and fan, but one joint is necessary in fixing these closets."

The simple U-shaped overflow enters the receiver, so as to face the valve when it is open. In the more costly closets of this kind are silvered-glass valves, hermetically sealed, so that a bright surface is seen on looking into the basin; also white, gold-lined, and fancy basins, cut-glass and ivory handles, with silver-plated dish and fans. This closet has been extensively used in England. In the illustration, one of Underhay's "air-valve regulators is shown in position attached to the closet. This regulator will be described in the last section of the book.

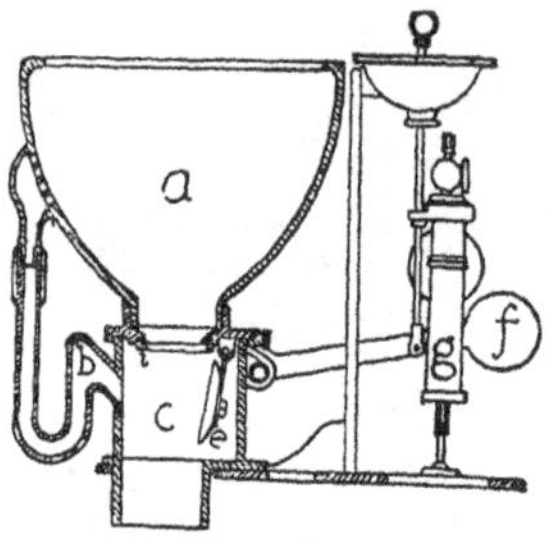

Fig. 72.
Underhay's Valve Closet.
a, Bowl. *b* Overflow. *c*, Receiver. *e* Valve. *f* Weighted lever. *g* Brass regulator-valve. *i*, Ring of India-rubber.

Demarest's Whirlpool, Climax, and Acme.—Mr. John Demarest, of the firm of J. L. Mott & Co., New York, has been a prolific inventor of water-closet apparatus, and his closets have been exten-

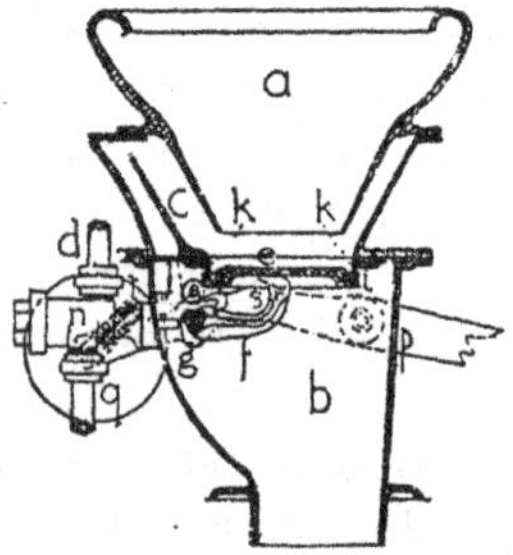

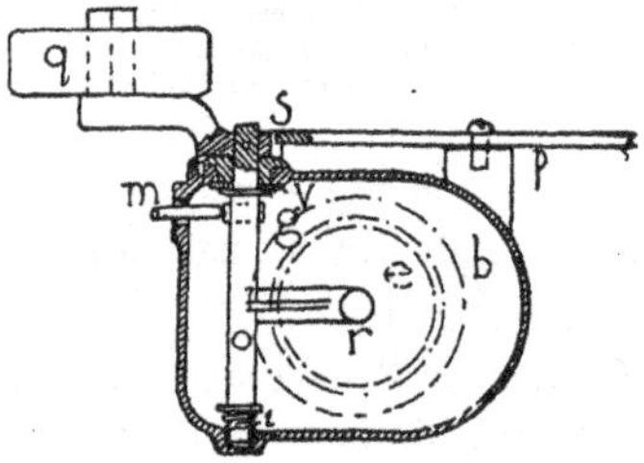

Demarest's Closets.
Fig. 73.—"Whirlpool" Closet. Fig. 74.—Section of Receiver.
(See Fig. 76 for explanation of reference letters).

sively used in this country. He received letters patent from the United States for a valve-closet, and also from Great Britain for the same closet in 1876. J. L. Mott & Co. manufacture this closet under three names: "Whirlpool," "Climax," and "Acme." The valve and its mode of working and the receiver are the same in each case. In the "whirlpool" closet the bowl is placed in a metal hopper that is joined to and forms a part of the receiver or container, while in both the other closets the bowl is placed directly on the receiver, and held in place by metal buttons or strips bolted to the receiver, and projecting over a flange at the bottom of the bowl. In the "Acme" closet the bowl is attached to the receiver in the usual manner by set-screws. I quote the following from Demarest's specifications:—

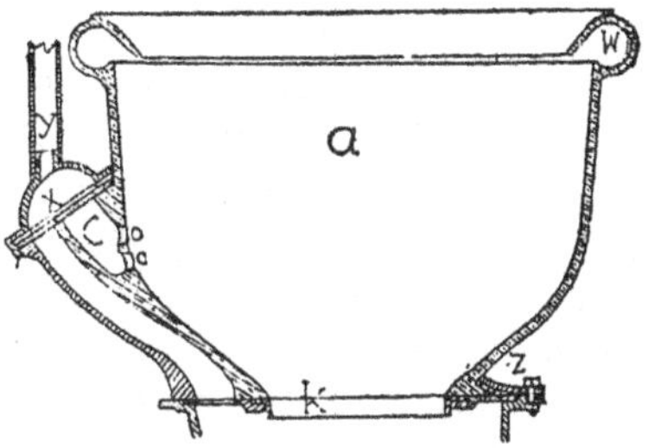

Fig. 75.—"Climax" Closet.

"The basin-valve is operated by an arm upon a weighted rock-shaft; the opening through which the shaft (or spindle) passes is rendered tight by a washer. The weighted arm of the rock-shaft is operated upon by teeth upon a lever which receives motion from an ordinary water-closet pull. The valve which admits water to the bowl is operated by a cam on the rock-shaft, and is provided with a cup leather piston and a spring to regulate the gradual closing of the spring.... The valve (at the bottom of bowl) is made of an elastic ring below the porcelain or enamelled surface. The weight on crank arm is preferably provided with a spring between the weight and its arm, to prevent concusssion."

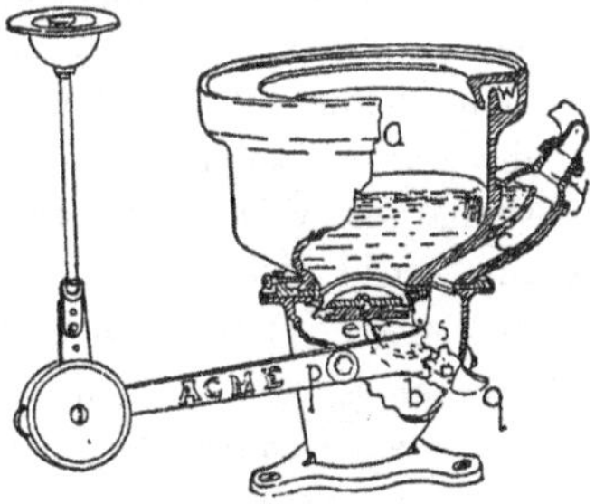

Fig. 76.
Perspective View.—"Acme" Closet.

Demarest's Valve-Closets.
(Figs. 73, 74, 75 and 76.)

a, Bowl. *b*, Receiver. *c*, Overflow. *d*, Supply-pipe. *e*, Valve. *f*, Finger, or lever to support valve. *g*. Cam to open supply-valve. *i*, Rubber-ring. *k*, Metal-seat. *l*. Swivel. *n*. Supply-valve. *m*, Rod to operate supply-valve. *o*, Rock-shaft, or spindle. *p*, Weighted lever. *q* Counterpoise. *s*, Toothed gearing. *r*, Spherical end of finger. *t*, Springs. *v*, Washer. *w*, Flushing-rim. *x*, Flap-valve. *y*, Vent to overflow.

In practice the weight is not attached directly to the spindle, as described in the specifications, but is placed at the end of the lever, which is connected with the hand-pull. (Figure 76).

The overflow in the "Whirlpool" closet is between the bowl and the hopper in which the bowl is set; while the "Climax" and "Acme" have the greater part of the overflow moulded on the bowl. A vent pipe enters the overflow at its crown, and is intended to carry off any gases generated in the receiver and prevent the trap of the overflow from being siphoned. The overflow of the "Climax closet" has a flap-valve as an additional protection against gases entering the room through the overflow. This flap-valve was probably added because of Dr. Fergus's experiments on the permeability of water-traps; while Dr. Carmichael's experiments on the same subject at a later date have caused the manufacturers to leave it out of the "Acme" closet. The "Climax" and "Acme" closets both have flushing-rims which differ from each other in their form.

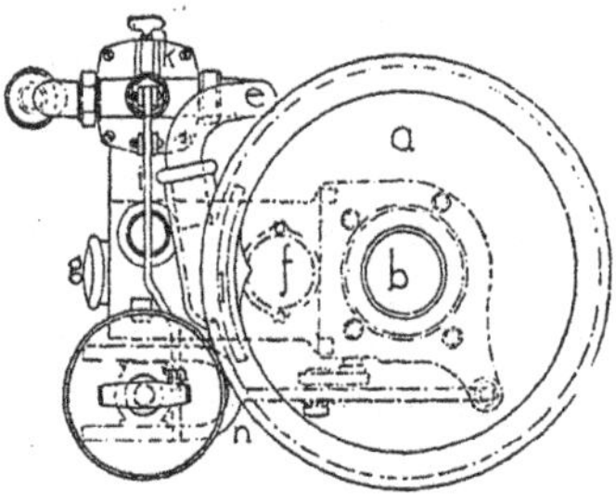

Fig. 77.
Top View.—Jennings's Valve Closet.
a Bowl. *b*, Valve. *c* Receiver. *d*, Vent-pipe. *e*, Supply. *f* Ball-valve. *g*, Opening for lavatory waste. *h*, Overflow. *i*, Lever for supply-valve. *k*, Waste-preventer. *l*, Fan.

Jennings's Valve-Closet. — J. G. Jennings, of London, invented a valve-closet which was patented in this country in 1880, having previously received letters patent from Great Britain in 1878. This closet Jennings claims as an improvement on the Bramah. The bowl supply, receiver, vent-pipes, and manner of working the valve have nothing new in them; the principal novelty is in the arrangement of the overflow. The valve was invented by Jennings in 1868. The overflow empties directly into a U or half-S trap, in which there is a bell opening for a lavatory or bath waste (this is claimed as a novel feature). There is a ball-valve in the overflow which has its seat on a sharp metal ring. This ball would take and keep its seat by the action of gravity, being raised only by water coming into the receiver, and backward pressure either of water or gas would only tend to make the ball fit more tightly, unless some foreign substance should get between the ball and its seat. To prevent waste matter entering the overflow, the valve at the bottom of the bowl, when open, closes the outlet of the overflow into the receiver. The receiver is properly vented.

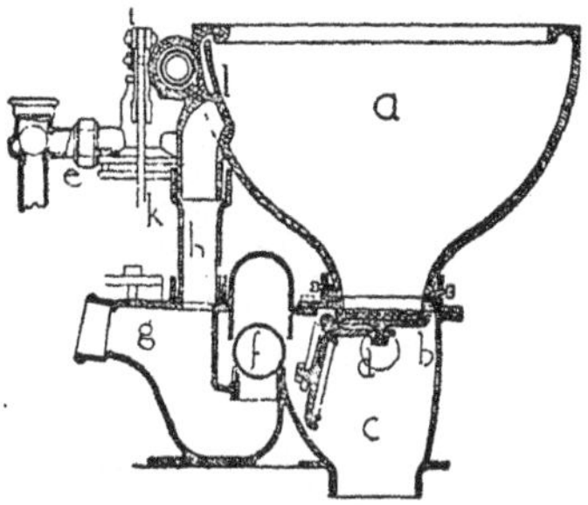

Fig. 78.
Section.—Jennings's Valve-Closet.

In 1868 Jennings received patents for an improvement on valves for closets of this class. A vulcanized India-rubber band was properly stretched in a groove or slot which was made in the valve for this purpose. In a fitting attached to the receiver was another vulcanized-rubber ring of the same diameter as the one in the valve, the bottom of the last rubber ring being V-shaped. When these two rings come in contact, as they would do when the valve is closed, in a valve of this kind, the joint would remain water-tight,

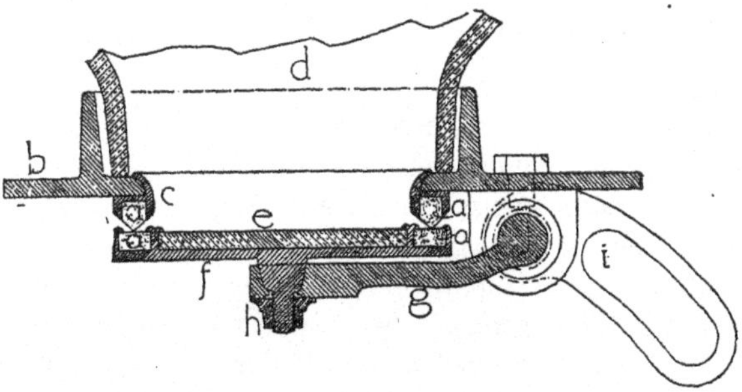

Fig. 79.—Section.—Jennings's Valve.—Details.
aa, Vulcanized-rubber rings. *b*, Top of receiver. *c*, Collar, or fitting for rubber ring. *d*, Bowl. *e*, Porcelain face. *f*, Metal back. *g*, Arm. *h*, Nut. *i*, Slotted crank. *j*, Spindle.

even in case a small foreign body should come between the points of contact with its seat. The efficacy of this joint and many others described depends upon the durability of vulcanized rubber. This valve combines the best points of all the valves belonging to the useful closets of this class. It will be seen by reference to the dif-

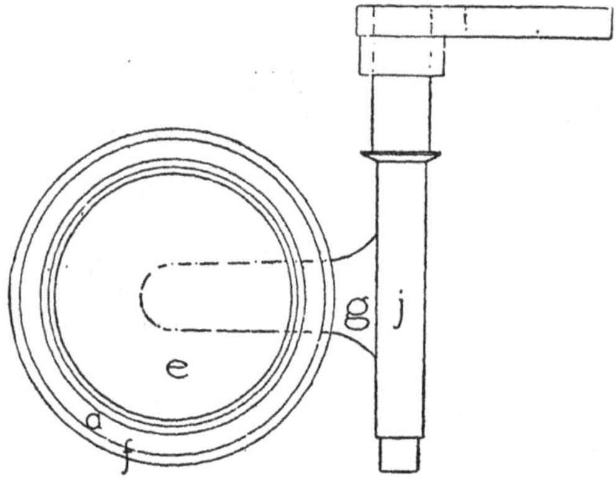

Fig. 80.—Top View.—Jennings's Valve.

ferent closets described herein, that in the practicable closets the valve either has a rubber disc or ring around its circumference, or has a seat against a rubber ring, while this closet has the ring in both places, one being a portion of the valve, and another forming the valve-seat.

The Alexander Closet.—The Alexander closet was invented in 1880, and is called by the inventor the "Sanitary Closet." This is one of the few valve-closets that have a ball-cock and compartments for floating it connected with the bowl. In 1835 John Odys combined a float supply-valve-chamber with a valve-closet. The patentee of the Alexander closet claims that "there is no possibility of the float being fouled; the bottom of bowl and lower valve being thoroughly washed each time by a flush from tank" (ball-cock chamber). The top of the valve, which is porcelain, can be taken off, and a new washer placed on without d'sconnecting the closet. There is a connection between the supply-chamber and valve-seat, the flowing of water through which is supposed to prevent paper, etc., from lodging either on the valve or its seat. The rock-shaft or spindle has a stuffing-box to trunk to prevent an escape of gases generated in the receiver. The height of water in the bowl may be regulated by adjusting the float. The overflow is

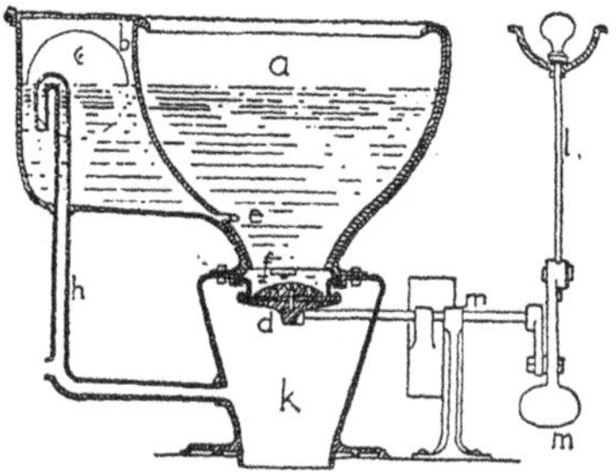

Fig. 81.

Alexander's Closet.

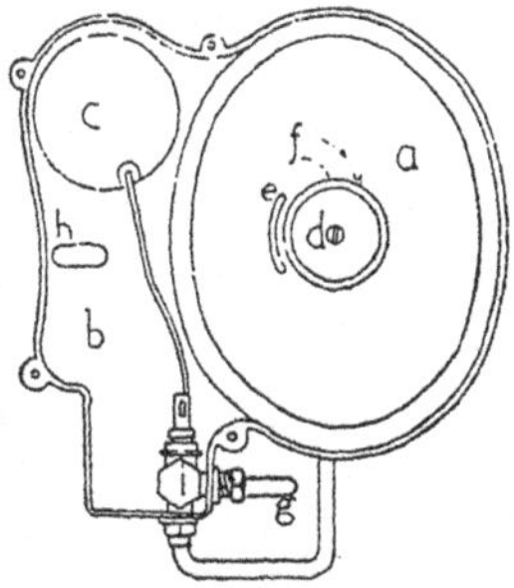

Fig. 82.

a, Bowl. *b*, Supply-valve compartment. *c*, Float for supply-valve. *d*, Valve. *e*, Opening between bowl and supply-valve compartment. *f*, Opening to wash valve. *g*, Water-supply pipe. *h*, Overflow. *i* Supply-valve. *k*, Receiver. *l*, Hand-pull. *n*, Weights on rock-shafts.

"doubly-trapped and has an outlet for ventilation." These closets are furnished by Miller & Coates, of New York, made entirely of earthenware, or with an iron body in which the earthenware bowl is set; the side of the bowl forming, in this manner, one side of the supply-valve chamber.

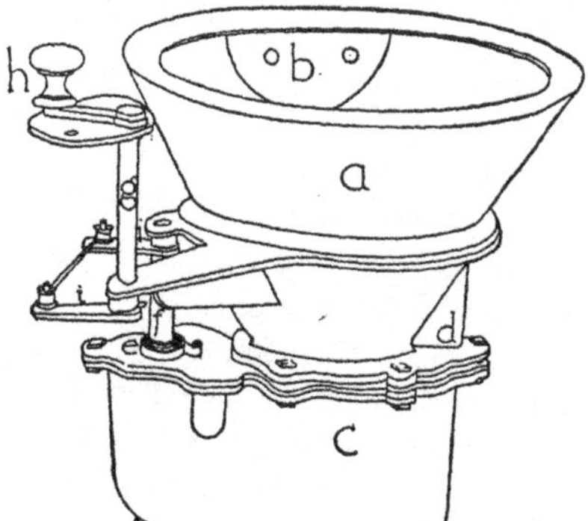

Fig. 83.—Perspective.—Llewellyn's Closet.

Llewellyn's Closet.—In 1864, Peter Llewellyn, a Welshman, invented a closet belonging to this class in which the valve is worked by means of an eccentric inclined plane or cam that turns on a vertical axis. Motion is imparted to this axis either by means of a crank placed just above and parallel to the seat, or by means of a hand-pull connected by a combination of wheels with this upright axis. The valve is kept in position by means of a spring which encircles the axis. The arm to which the valve is attached has a projection or stud that rests on the cam, which, when the axis is turned either by the crank or hand-pull, travels or slides up or down the surface of the inclined plane, at the same time causing the valve to open or close. The valve is of metal and it has a seat of a more pliable material. This closet has a trapped overflow. The different parts of the closet and their arrangement are clearly shown in the illustrations.

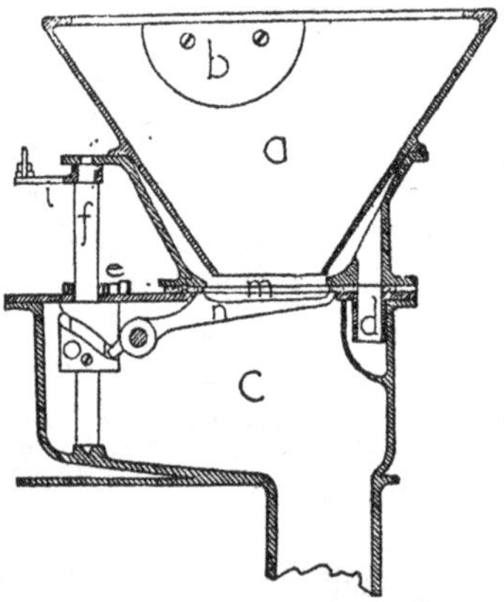

Fig. 84.—Section.—Llewellyn's Closet.

a, Bowl. *b*, Fan. *c*, Receiver. *d*, Overflow. *e*, Spring. *f*, Vertical axis. *g*, Connecting-rod. *h*, Knob for hand. *i*, Connecting-levers. *m*, Valve-seat. *n*, Valve. *o*, Cam or inclined plane.

Preston's Closet.—There was a closet of this class invented in England by one Preston in 1876. The novelty consisted in three valves, each in its separate receiver, one directly below the other, and so arranged that when the top valve is open the second one is closed. In this way it is intended to shut off direct communication with the sewer.

Rice & Sargent's Closet.—A closet was invented in this country in 1876 by Rice & Sargent. This closet was intended to be placed below the point of discharge. The spindle on which the valve works, and to which it is connected by an arm six or eight inches long, is turned by the same lever and at the same time that a piston, which fits into the cylinder below the receiver is pressed down. The waste matter is forced into the soil-pipe by the action of this piston. There is a valve at the entrance to the soil-pipe that opens only in a downward direction, so as to prevent the return of waste matter into the cylinder. This valve is held in position by a spiral spring. The mechanism of this closet does not appear to be as simple as that in the "Bunnet" or "Sand" closets (Figs. 47, 48, 51–52) both of which are intended for use below the water line. Among other closets of this class, in which the trap, when one is used, is below the floor-level, I will mention the "Victor" closet, in which the valve is moved by a toothed lever, similar to the Demarest valve-closets; Bolding's simple closet; Atwater's closet; Edwards's closet, worked by geared tooth wheels; Blackwood's closets, for he has invented several complicated ones belonging to the above class; James & Drewett's closet, similar to the Bramah, differing from it only in the combination of the levers for opening the valve. The closets just mentioned above are either in use at the present time or have been recently invented.

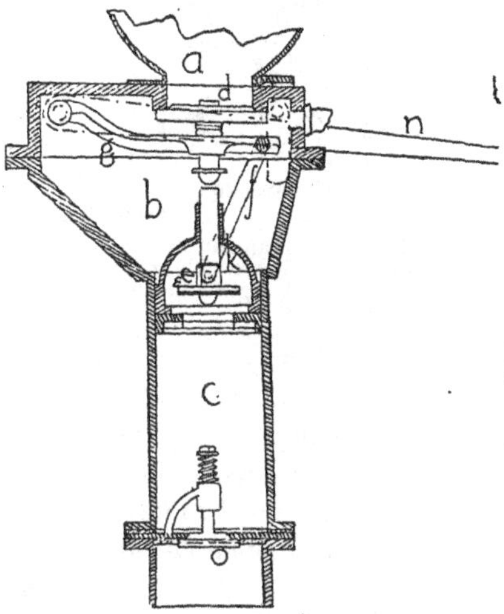

Fig. 85.—Rice & Sargent's Closet.
a, Bowl. *b*, Receiver. *c*, Piston-chamber. *d*, Main valve. *e*, Piston-valve. *f*, Connecting-bar. *g*, Valve-arm. *k*, Piston. *o*, Valve at entrance to Soil-pipe.

Other closets, which properly belong to this class and type have been patented and manufactured, but the variation is so slight from some of the closets that have been already described, that I do not think it necessary to describe them.

Tylor's Valve-Closet.—Two prominent English firms manufacture valve-closets in which the valve-chamber or receiver is formed in one piece with the trap, the combined trap or receiver being placed above the floor-level. J. Tylor & Son, London, claim the trap above the floor as a novelty in their valve-closet, which was invented in 1874, and improved in 1876–1878. The bowl is connected with the receiver by small set-screws. The space between the bowl and the flange is filled with putty or a cement of white and red lead or some similar composition. From the different

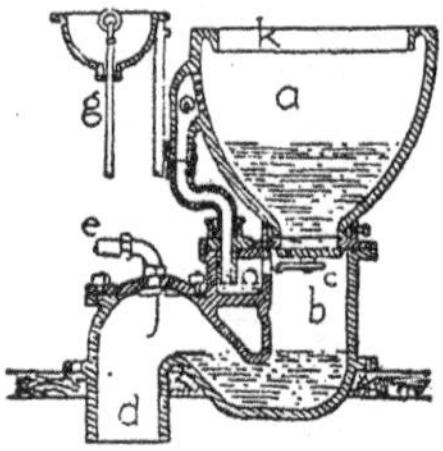

Fig. 86.—Section.

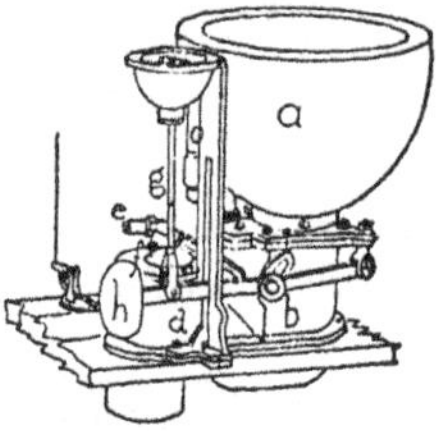

Fig. 87.—Perspective.

Tylor's Valve-Closets.

a, Bowl. *b*, Trap and Receiver. *c*, Valve. *d*, Entrance to soil-pipe. *e*, Vent-pipe. *f*, Removable cover for hand-hole. *g*, Hand-pull. *h*, Weighted lever. *o*, Overflow.

illustrations it may be seen that this is a common mode of connecting the bowl and receiver. The receiver is made of galvanized, or, more properly speaking, zinc-coated iron, and the valve, when open, closes the mouth of the overflow trap, thus, the inventor claims, keeping filth from entering the overflow, and at the same time preventing siphon action from taking place. The trap has a small box cast on the part back of the valve, into which the overflow pipe runs, and dipping below the water-line it forms a trap.

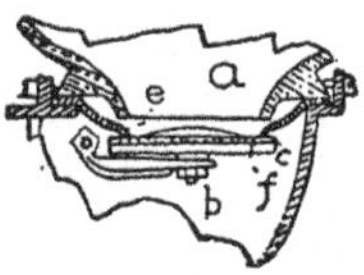

Fig. 88.

Detail of Valve.—Tylor's Closet.

a, Bowl. *b*, Receiver. *c*, Valve. *e*, Metal-seat. *f*, Leather or rubber.

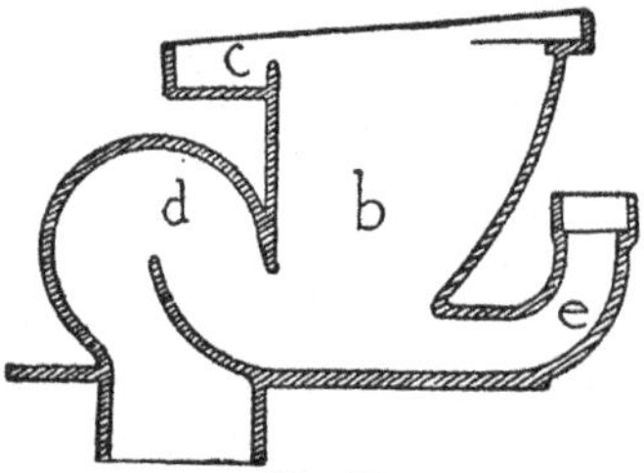

Fig. 89.

Improved Receiver.—Tylor's Closet.

b, Receiver. *c*, Box for overflow. *d*, Trap. *e*, Opening for bath or lavatory waste.

Tylor & Son manufacture several valve-closets which differ from each other only in unimportant details. They invented, in 1878, as an improvement on their old closet, an inlet into the receiver running below the water-line, for a bath-tub or lavatory waste-pipe. We have all probably had experience in the effect produced when a trap like this becomes accidentally stopped in any way, as this was a common mode of connecting the waste-pipes (it saved a

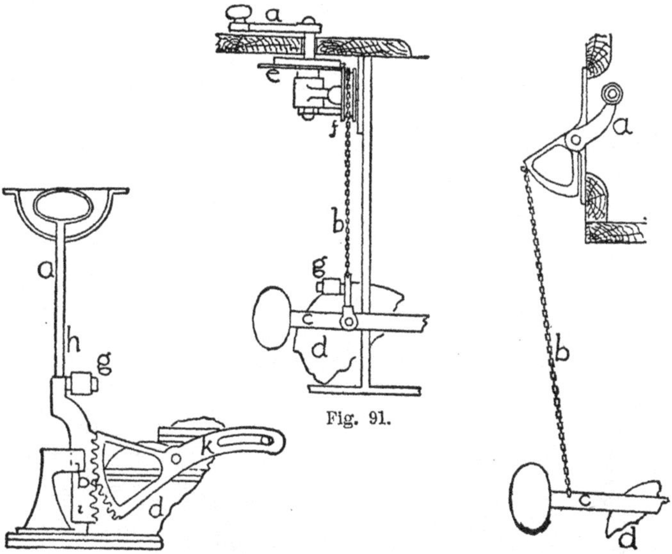

Fig. 90. Fig. 91. Fig. 92

Details of Tylor's Valve-Closet.

a, Crank. *b*, Chain. *c*, Lever. *d*, Trap to closet. *e* and *f*, Wheels at right-angles. *g*, Stud for supply-lever. *h*, Hand-pull. *i*, Rack. *k*, Toothed-lever.

trap) a few years ago in this country. A small number of plumbers still continue the bad practice.

In the closet under consideration the valve is very similar in construction to the Hellyer valve, a rubber or leather disc enclosed between an earthenware disc on top and a metal one beneath. The earthenware disc is the smallest, thus enabling the pliable disc to fit against a projecting metal seat. The traps of these closets have an inspection cover bolted to the trap, with a vent-pipe inserted into the crown; the vent is so very small as to be almost, if not quite, useless. There is no vent-pipe to the receiver. Tylor claims as his invention the three ways of opening the valve illustrated on page 57. The first method was by forming a rack and pinion of the hand-pull and a segmental projection on the lever that opened the valve. The second method was by means of a crank placed in a horizontal position above the seat (see Fig. 91), as shown in the cut. This crank when turned imparts motion to a combination of wheels, on one of which is a drum upon which a small chain is wound (the drum is in a vertical position). In this manner the weighted lever is raised or lowered. The third method consists in simply attaching a chain to the lever and a bell-crank; any pressure on the crank would raise the lever.

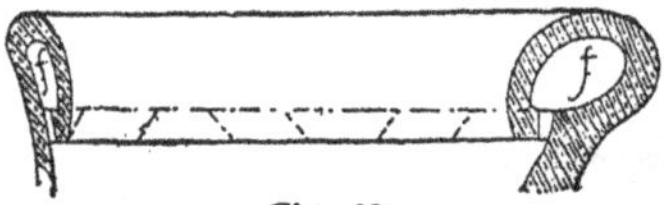

Fig. 93.
Flushing-Rim.—Tylor's Closet.

Doulton Lambeth Valve-Closet.—Doulton & Co., of England, manufacture a valve-closet in which the receiver or valve-compartment and trap are in one piece, and this piece is placed above the

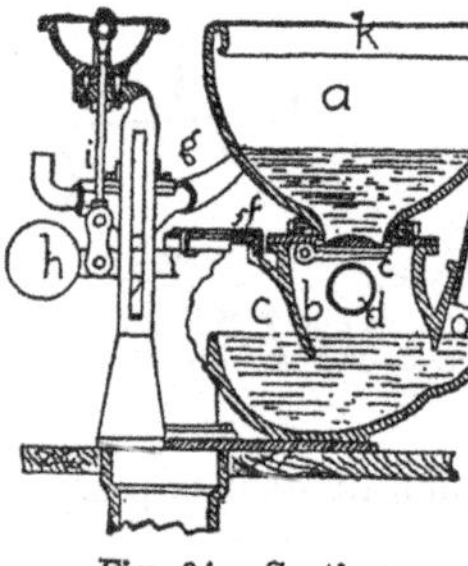

Fig. 94.—Section.

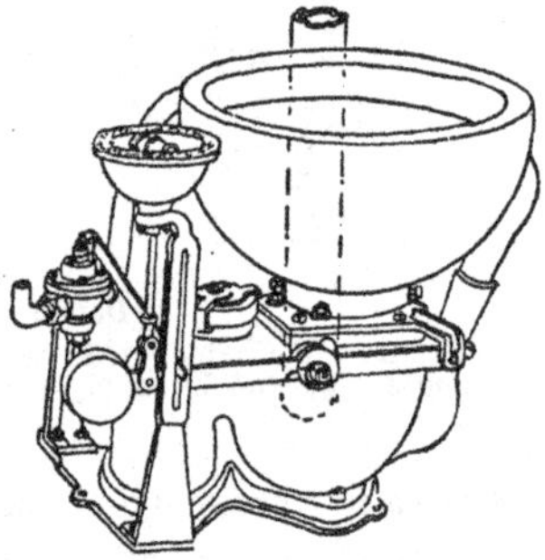

Fig. 95.—Perspective.

Doulton Lambeth Valve-Closet.

a, Bowl. *b*, *c*, Receiver and trap. *d*, vent-pipe. *g* Supply-pipe. *f*, Inspection-hole. *h*, Weighted lever. *i*, Supply-valve. *k*, Flushing-rim. *o*, Overflow.

floor. This closet appears simple in its mechanism. The manner of trapping the overflow by introducing it into the main trap below the water-line is a novelty with this class of closets, and while it has the advantage of insuring a water-seal for the overflow, it is liable to become foul where it enters the trap. The receiver has an ample vent-pipe, but the trap is not ventilated from the side where it is connected with the soil-pipe, as it should be. "The trap is provided with an opening for inspection, ... which is secured with a patent cap ground into the opening, and so constructed that a slight turn one way or the other disengages or fixes it. .... The valve has an earthenware facing, and the interior of the trap is lined with strong, smooth glaze, perfectly impervious and incapable of corrosion," for these reasons offering the least resistance to water or soil passing through it, and at the same time insuring more perfect cleanliness. The trap is furnished so the bowl may be placed in front, or on the right or left side, of the soil-pipe. In the perspective, the mode of attaching the Doulton patent supply-valve is shown.

DEDUCTIONS.

If it should be necessary at any time to select a valve-closet, one should be chosen with simple mechanism and few parts. The only closets of this class which appear to be practicable are those with hinged valves, the valves having either a rubber seat or disc, or both seat and disc of rubber. The time the valve will last without fixing depends on the durability of vulcanized rubber. The receiver should always be enamelled, earthenware being still better, as there is a possibility of the enamel being thrown off by corrosion or rust between the enamel and its iron back. The receiver should be ventilated, as there will always be more or less offensive matter deposited on its sides; for this reason the smaller and smoother the receiver is the better. The vent-pipe in the receiver also keeps the overflow-trap from being siphoned by the water discharged from the bowl. This compartment should also be separately flushed, although I have no idea that a surface, no matter how well glazed or vitrified, over which fæcal matter, urine, and water in which these matters are floating in particles so finely divided as to be invisible, can be kept clean without the application of soap and water on a mop or swab by hand. I consider the variety in which the trap is above the floor, forming at the same time the receiver, as the best arrangement for this part of the closet, but the advantage gained would not counterbalance lack of ventilation.

I consider the side-outlet valve-closet, in which the valve opens in an outward direction from the bowl, as the best of this class. The back of the valve never coming in contact with the waste matter, the only part of the closet between the bowl and the trap that could become foul and not be noticod would be the short pipe between the valve and the floor; while this would become more or less foul, it would be scoured as thoroughly as a column of water unaided could wash any surface; but a trap must be used below the floor. These closets are far superior to the pan-closets, of which I will give a description in the following pages, in their capacity for retaining a large amount of wat r in the bowl, and in the receiver being smaller; but they are defective in having a compartment, however small, between the bowl and trap when they have a trap, and the valves are liable to leak when least expected; then the advantage of the body of water in the bowl is lost. The overflow, unless filled by a driblet, will be emptied by evaporation. It will have been noticed in the foregoing descriptions that the better closets of this class have a flushing-rim, by which water is supplied to all parts of the bowl at the same time from a cistern. Where these closets have only a fan supplied from a supply-valve, the waste matter is liable to remain in the trap until the bowl has been emptied several times.

:horities agree, without an exception, that pan-closets should never be used; but as they have been so extensively introduced in all parts of the civilized world, their description and history properly form a part of the literature of the subject which is under consideration.

I think the pan-closet was in use before a valve-closet was invented, as I do not find a single instance in the specifications belonging to the Patent Reports in which the pan is claimed as a novelty; on the contrary, the pan is always mentioned as if it

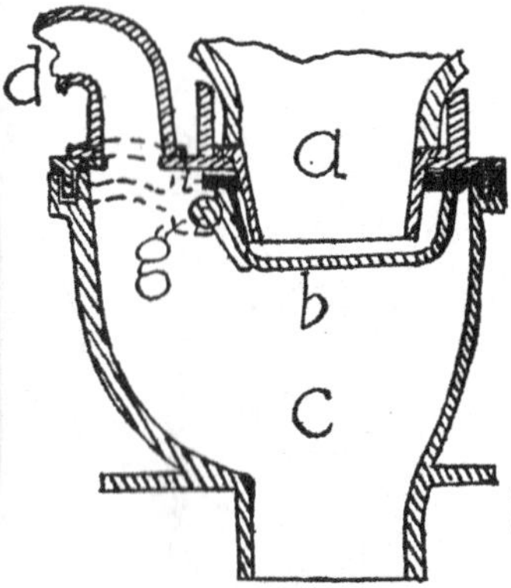

Fig. 98.—Section.
Details of Tylor's Pan-Closet.
a Bowl. *b*, Pan. *c*, Receiver or container. *d*. Overflow. *f*, Leather seat. *g*, Spindle.

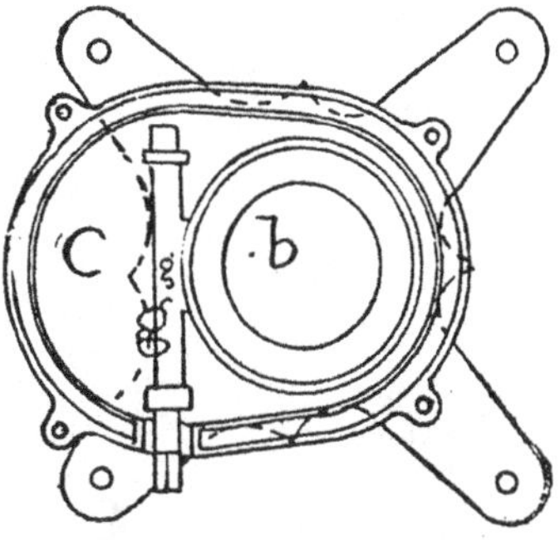

Fig. 99.—Container.—Plan.

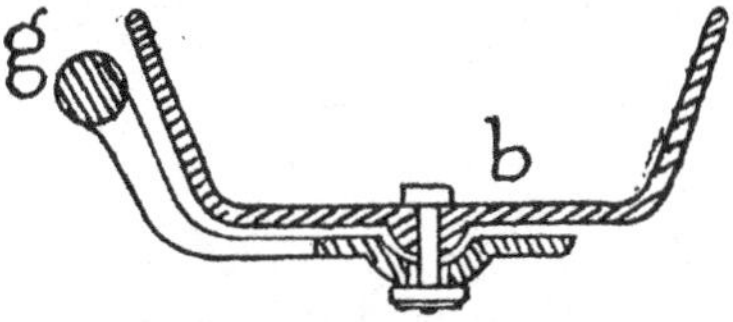

Fig. 100.—Pan.—Tylor's Closet.

were so well known that it would be unnecessary to describe it. This class of water-closet is, or was, as I do not think any one would use a pan-closet at the present time, usually connected with a trap below the floor. In this country the trap was usually a lead or iron siphon-trap, while in England one "container" of filth was added to a second filth accumulator, by joining the receiver to a D-trap.

The following descriptions will show how persistent have been the

SECTION III.

PAN-CLOSETS.

ACCORDING to the definition given on a pre pan-closets include all that have a dish-sha pan at the bottom of the bowl. The pan is hold a sufficient quantity of water to allow the bottom or a conical attachment to the top of the receiver (its uppermost), to dip into it below the water-line; in thi

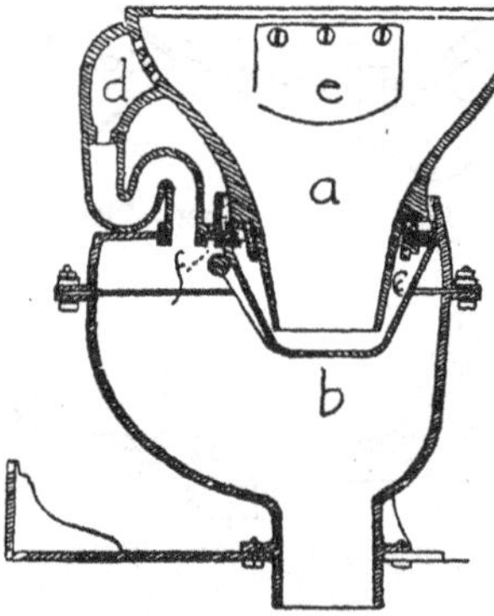

Fig. 96.—Section.

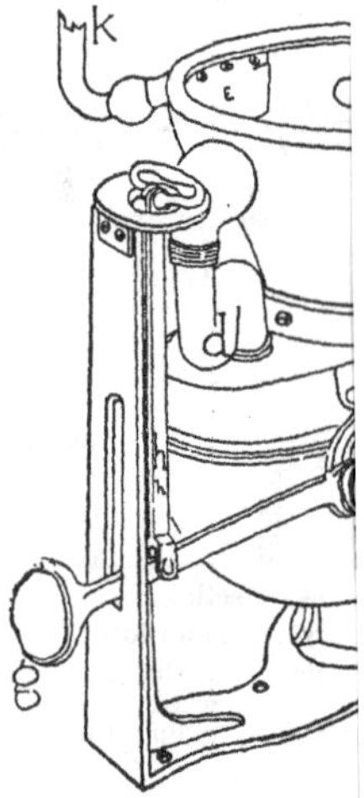

Fig. 97.—Perspec

Tylor's Pan-Closet.

a, Bowl. *b*, Receiver. *c*, Pan. *d*, Overflow. *e*, Fan. *f*, Leather
g, Weighted lever. *h*, Hand-pull. *i*, Slotted crank. *k*,
m, Set-screws.

water-seal is formed. Except in rare instances the pan i to form the overflow. The receiver in this class of closet sarily large, and filth accumulates and remains in it. Sa efforts of inventors and manufacturers to obtain some contrivance by which the container of a pan-closet might be kept clean and wholesome.

In the different trade-catalogues, pan-closets usually have the title "valve-closet" prefixed to them. This term, when applied to the pan, is a misnomer, but the term usually has reference to some form of supply-valve, and does not properly apply to the closet, as the supply-valve could be attached to other classes of closets with equal facility. Gases generated in the receiver may escape into the room when the pan is thrown back when opened, or through the hole in which the spindle works, as it frequently becomes loose. The bowl being generally placed on the receiver, a joint of putty or white and red lead cement is depended upon to keep it in position. This putty is generally full of cracks, the bowl being loose, thus allowing access for gas. The container of pan-closets requires an air-hole, and Hellyer tells us that in England this air-hole is usually left open and allows the smell to enter the room direct. There is little doubt that the days of pan-closets are numbered, and that in a few years, at most, we will never see them put into even the cheapest house. Pan-closets may be properly divided into two types, one in which the pan fits tightly against the bowl or projection, and the other and most numerous type where the pan is only intended to form a barrier to sewer-air by forming a water-seal.

Tylor's Pan-Closet.—The first mention of a pan-closet in which the pan was intended to form a water-tight joint was made by Edward Tylor, of London, in 1829. In the specifications he says: "this is similar to what is known as a pan-closet," and he claims as his invention the application of leather or other suitable material against which the edges of the pan or saucer (as he calls it) might have a seat and form a water-tight joint. Further on in the specification he states, näively, that should the joint leak, even then it would be in every way equal to a pan-closet. In this closet is illustrated one of the early instances of a weighted lever. Modifications of this weighted lever have been in general use ever since, on valve and pan closets, to hold the valve or pan in position. The container and pan were made in different sizes and shapes, as is shown clearly by the illustrations that accompanied the specifications.

In this closet is seen one of the few instances, in closets of this class, of a separate overflow for the bowl; the pan fitting tightly makes it necessary. The bowl is fitted into place more securely than is usually the case with closets of this class, being set into a metal rim or collar, and held in position by small set-screws. The pan in some cases was attached to an arm that was joined to the spindle. By removing the bolt in the centre of the pan, the pan could be taken out without removing the spindle. The overflow would probably be siphoned by the sudden discharge of waste matter into the receiver.

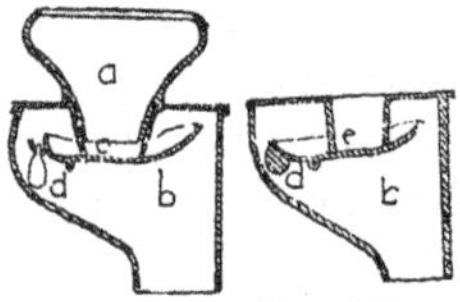

Fig. 101. Fig. 102.
Rogier & Mothe's Closet.
a, Bowl. b, Receiver.
c, Pan. d, Weight.

Rogier & Mothe's Pan-Closet.—In France this class of closets has been in common use for years, and in some cases it seems to have been the intention to have the pan fit tightly against a metal rim. The device of Rogier & Mothe illustrated by Liger in his work, has a pan balanced by a weight on the side of the axis opposite to the pan. When a certain amount of water had accumulated in the bowl, the pan would tilt and drop its contents into the receiver. In one case the weight forms a part of the pan; in the other instance the weight had an eye, to be placed over a hook on the rim of the pan. This closet is very much like Flament's closet in the manner of balancing the pan in the first instance and the valve in the second instance. This closet received a first-class medal at the Universal Exposition held in Paris in 1855. In the larger number of closets belonging to this class the pan is made fairly deep, and is only for the purpose of forming a water-seal at or near the bottom of the bowl,

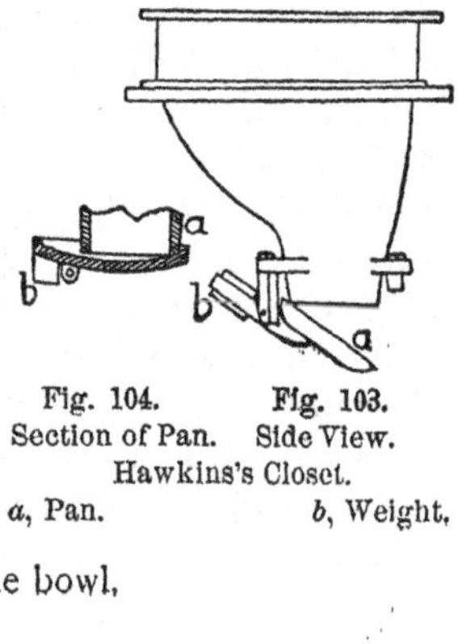

Fig. 104. Fig. 103.
Section of Pan. Side View.
Hawkins's Closet.
a, Pan. b, Weight.

Hawkins's Closet.—In 1821, Stephen Hawkins, an Englishman, received letters-patent for a pan-closet on the same principle as the one just described, except in the latter case there was no intention of having the pan fit tightly against the bottom of the bowl. The weight is attached directly to the end of the pan, as in the case of Rogier & Mothe's closet, or it forms one end of a balanced lever, the pan forming the other end. These forms are shown in Figures 103 and 104.

Beacham's Pan.—In England, in 1825, Beacham invented a curious arrangement for holding a pan in position. This consisted of a band-spring enclosed in a cylindrical box; to this box was attached a short arm. On the end of the arm is a small wheel. This wheel pressed against and rolled on a hinged bar, to which the pan was joined. When the pan is pressed down by a lever or crank in the usual manner, the wheel would roll along the bar, the short arm would be depressed, and the spring wound up more tightly than it was. When the pressure was removed from

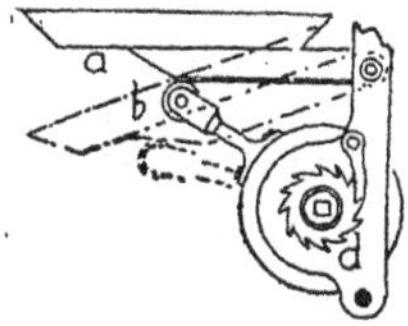

Fig. 105.—Side View.

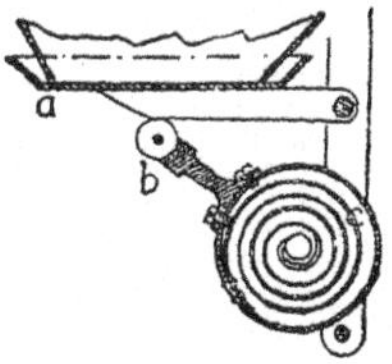

Fig. 106.—Section.

Beacham's Pan-Closet.

a, Pan. *b*, Wheel at end of arm. *c*, Spring in and attached to box. *d*, Rachet and catch.

the lever, a contrary movement would take place, and the pan would be forced into position again by the spring. If the spring should become weak, it could be tightened by a wrench to any desired tension, a rachet wheel and catch being placed there for the purpose. This pan was intended for use either on commodes or water-closets, but any one can see at a glance there is too much mechanism in the receiver for practical purposes.

Findon's Pan-Closet.—In 1836, James Findon, an Englishman, invented a combination of levers and arms for opening a pan-closet. The pan is held in position by a long, weighted lever. A rod,

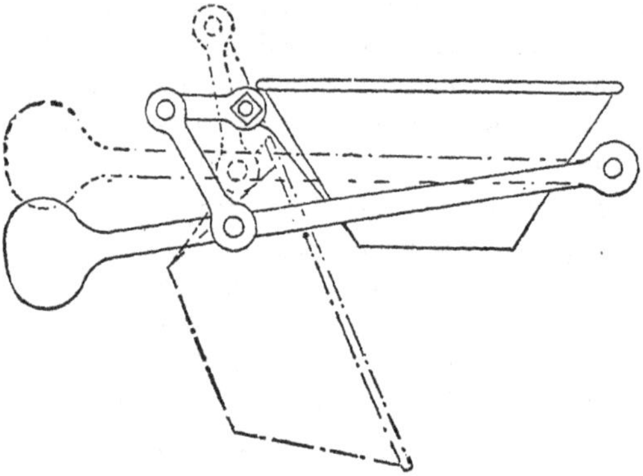

Fig. 107.—Findon's Pan-Closet.

which was connected with the lever near the weight, was also joined to a short lever that was attached to the pan. These arms and levers were so connected as to revolve freely for a certain distance on their joints, while the short arm or lever was connected rigidly with the pan. The spindle formed a fulcrum on which the pan would be balanced by the lever.

Havard's Closet.—In France the use of this class of closet seems

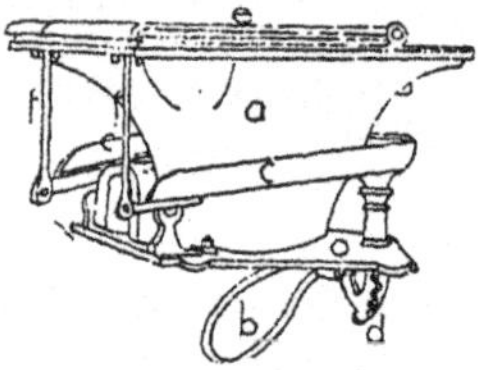

Fig. 108.
Perspective.

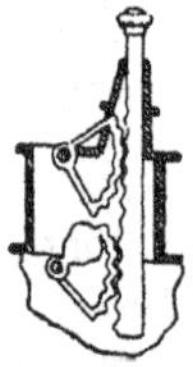

Fig. 109.
Section showing working parts.

Havard's Closet.

a, Bowl. *b*, Pan. *c*, Levers connected with rack. *d*, Toothed quadrant. *f*, Connecting-rods. *e*, Foot-rest.

to have been general. Among other French closets I note one invented in 1840 by Havard. This closet received a medal at the

Universal Exposition held in Paris in 1855. In this closet motion is imparted to the pan by means of a rack and mutilated pinion or toothed quadrant. The axis of one of the quadrants is the spindle on which the pan turns. Any pressure on the seat would cause the rack to partially revolve the quadrant that was connected with the pan. In this manner the pan would be opened whenever there was pressure on the seat. This closet was intended for use in public places, and Liger tells us that in 1875 it was still used in public places in Paris.

Guinier's Pan-Closet.—Guinier, a manufacturer of plumbers' supplies in Paris, in 1840 invented a closet which I illustrate as an example of a large receiver, and a complicated arrangement for accomplishing a simple movement. I must refer to the illustration to explain the manner of opening the pan. The pull-rod, levers, arms, connecting-rods, and a large, slotted shoe-shaped arrangement are all concealed in the receiver. The pull-rod passes through a stuffing-box. It will be readily seen that this machinery could not be repaired without taking the closet to pieces, and also that it would be liable to fail in a short time. Any arrangement of this kind would afford excellent opportunity for the accumulation of filth on the different pieces of machinery. There seems to be no effort made in any of the French closets to ventilate the receiver.

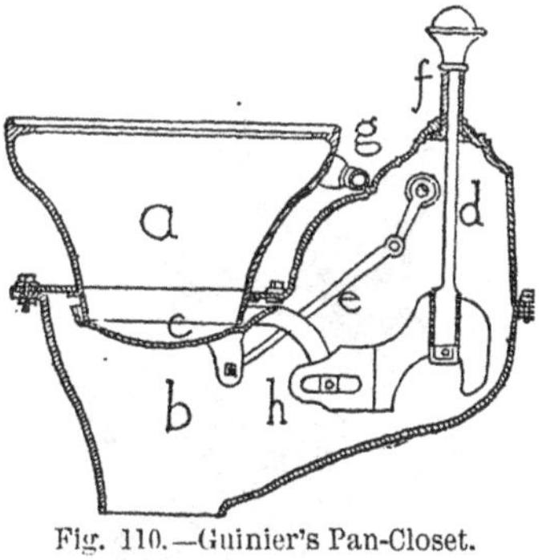

Fig. 110.—Guinier's Pan-Closet.
a, Bowl. *b*, Receiver. *c*, Pan. *d*, Pull-rod *e*, Connecting-rods. *f*, Stuffing-box. *g*, Supply-pipe. *h*, Slotted-shoe.

Common English Pan-Closet.—Baldwin Latham* describes and illustrates as types of pan-closets which had been in common use, and were still used in England in 1873, the closet of which the following is a description:—

* *Sanitary Engineering.* Sewerage, London: 1873.

In this closet are shown some of the worst features of the pan-closet. The bowl is conical in shape, fitting into a hole in the top of the receiver. The bowl is held in position by a projecting rim, moulded around it about half-way between the top and bottom; between this rim and the top of the receiver a layer of putty is usually placed. I have never seen a joint formed in this manner between the bowl and the receiver in which the joint was perfect unless it had just been made. The least pressure against the bowl or receiver, a jarring of the floor or supply-pipe, may cause the joint to open. The putty is generally full of cracks, the cracks containing putrid urine and other foul matter. It can be easily seen that all foul gases generated in the receiver can come directly into

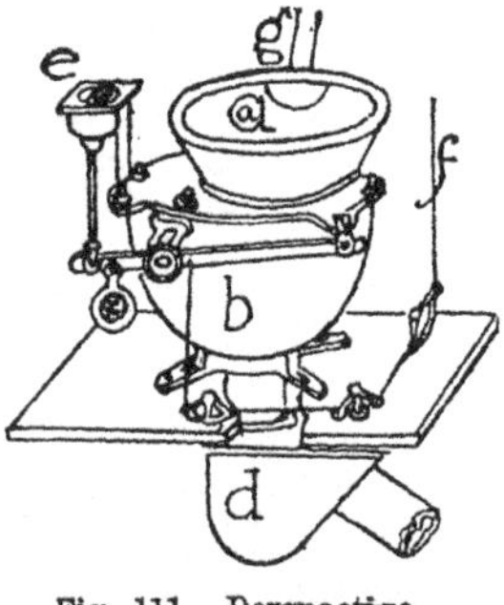

Fig. 111.—Perspective.

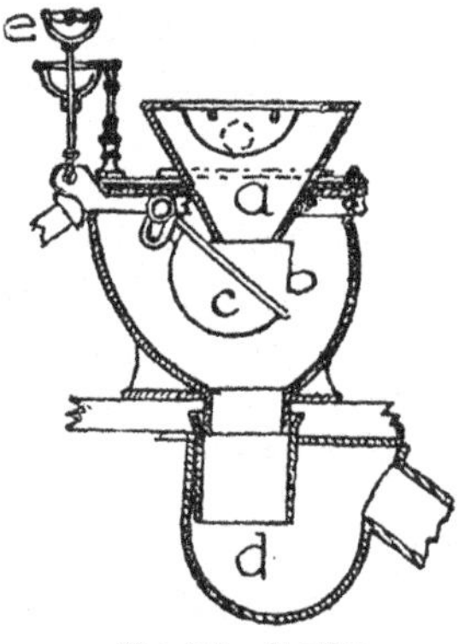

Fig. 112.—Section.

Common English Pan-Closet.

a, Bowl. *b*, Receiver. *c*, Pan. *d*, D-Trap. *e*, Hand-pull. *f*, Wires and bell-crank connecting with cistern. *g*, Supply-pipe.

the room when this joint is imperfect. The pan is moved in the usual manner by a spindle that has a slotted crank on one end connected with the hand-pull by a weighted lever. In the perspective sketch the mode of connecting the hand-pull with the cistern or tank by means of bell-cranks and wires is clearly shown. The usual receiver is shown in connection with a D-trap, both of which act as receptacles for the excrementitious matter passing through them. In these receptacles the waste matter lodges, remains, decays, and generates gases, and most probably liberates germs far more deleterious than would usually come from a well-ventilated sewer or drain.

The closets described in the following pages are manufactured at the present day by prominent manufacturers of plumbers' supplies. Patents, when claimed on them, are usually for some novelty in the position or form of the supply-valve, or the weighted lever. As these parts might be with equal facility, with some exception, attached to a valve-closet, they cannot be put down as distinct features in the class which is under discussion.

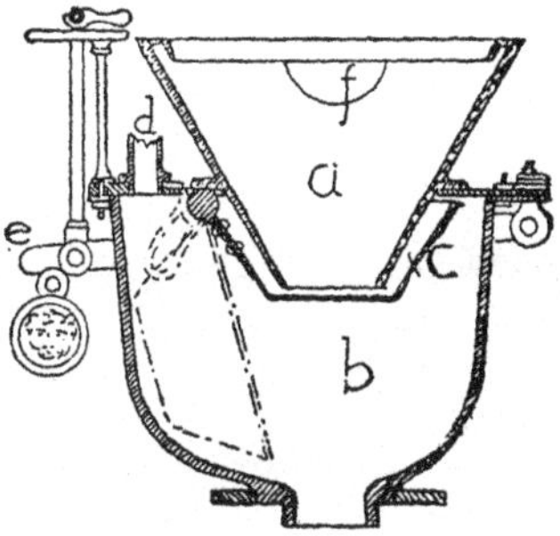

Fig. 113.—Hellyer's Pan-Closet.
a Bowl. *b*. Receiver. *c*, Pan. *d*, Vent. *e*, Weighted Lever. *f*, Fan.

Hellyer's Pan-Closet. — Although Hellyer* condemns the pan-closet in very strong terms in both his works on plumbing, he finds it a commercial necessity to manufacture them to supply the demand. This closet is similar to the one described above, except the vent-pipe, which is inserted into the top of the receiver. In this illustration is shown the usual manner of setting the bowl on the receiver, and the way in which it dips into the pan and forms a water-seal. This closet has a combination of the old-fashioned fan with a small flushing-rim for spreading the water around the bowl. The vent-pipe is screwed into the top of the receiver back of the pan, where most of the filth accumulates.

Doulton Pan-Closet.—The Doulton Company manufacture a pan-closet that has an ample vent-pipe which enters the receiver on the side near the bottom. This closet has a stoneware container, of which the manufacturer says: "and thereby is obviated the hitherto

* *The Plumber and Sanitary Houses.* S. S. Hellyer, London: 1880.
The Art and Science of Sanitary Plumbing. S. S. Hellyer, London: 1882.

great objection to this closet, from the container, when of iron, becoming encrusted with soil and emitting impure gases." This container has a white glaze inside.

Cazaubon's Closet.—F. Liger illustrates a closet invented by Cazaubon. In this closet the pan is moved by a short arm or lever, which moves up and down on a fixed axis. The end of the lever, where it comes in contact with the pan, has a semi-spherical button. The pan is hinged, and when it opens the end of the lever slides along the bottom of the pan. The principle is very much

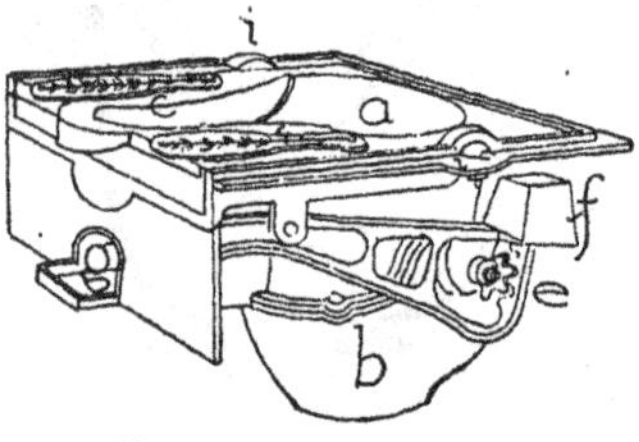

Fig. 114.—Perspective.

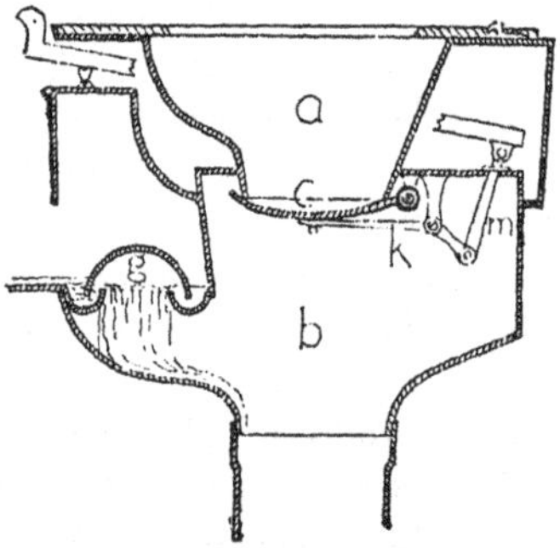

Fig. 115.—Section.

Cazaubon's Pan-Closet.

a, Bowl. *b*, Container. *c*, Pan. *e*. Cogs on lever. *f*, Weight on lever. *g*, Overflow. *i*, Journals. *k*, Short lever or finger. *m*, Connecting-rod.

like the one adopted by Mott in the Demarest valve-closets. The perspective shows an arrangement for opening the pan by pressure on the seat, or, more properly speaking, the stand, as the seat is arranged after the Oriental manner. The stand in this instance rests on two journals, and is connected with a lever, on the short arm of which is a rack. The rack imparts motion to a mutilated pinion, which, partially revolving on its centre, moves the short lever or finger that opens and shuts the pan.

Jones's Closet.—In the year 1860 a patent was issued in England to one Jones, for a tilting-pan very similar to the flushing-tanks now in use, which tilt when the water has attained a certain height. The pan was held in a level position by a small rod, which had a bearing on the rim at the back part of the pan. By raising this rod or hand-pull, the weight in the portion of the pan in front of the journals, on which it was balanced, would cause it to tilt and empty its contents into the receiver.

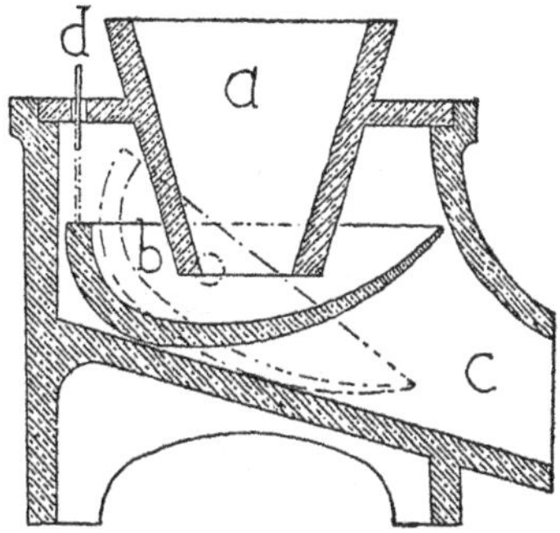

Fig. 116.—Jones's Pan-Closet.
a, Bowl. *b*, Pan. *c*, Receiver. *d*, Rod for holding pan in position.

Tylor's Pan-Closet.—Tylor & Son, of London, invented in 1878 a pan-closet receiver in which the trap and receiver were combined in one piece, both being above the floor. In this closet there is an inspection-hole in the crown of the trap, but no vent-pipe. The same firm manufacture an earthenware container, which is yellow on the the outside, and glazed with white on the inside. The top or cover to this container, and to which the mechanism is attached, is composed of galvanized iron; having the working parts connected with top of the receiver, the opportunity for breaking the earthenware portion is lessened.

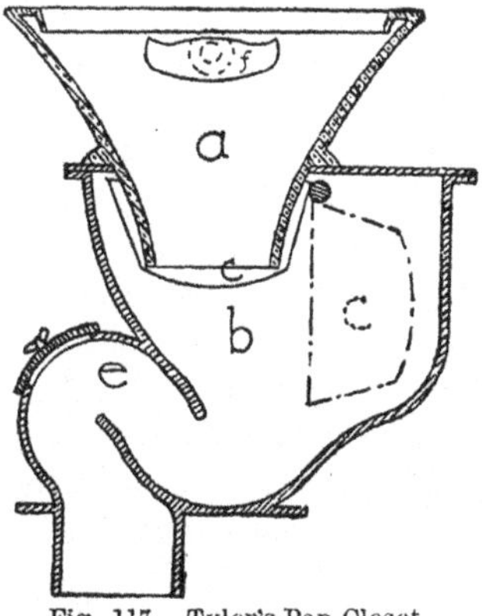

Fig. 117.—Tylor's Pan-Closet.
a, Bowl. *b*, Receiver. *c*, Pan. *e*, Trap. *f*, Fan.

These manufacturers also have a patent for a sprinkler to be connected with the container of a pan-closet. The sprinkler is divided into two branches, and is carried partially around just below the top of the container. In it are a row of small perforations. The sprinkler is connected with the supply-pipe, and when the water is turned on to flush the bowl, small jets of water are thrown against all sides of the receiver, from the different perforations. No doubt such a device would be of some service, although I have little confidence in an arrangement placed as this is, out of sight and out of mind. The pipe would corrode

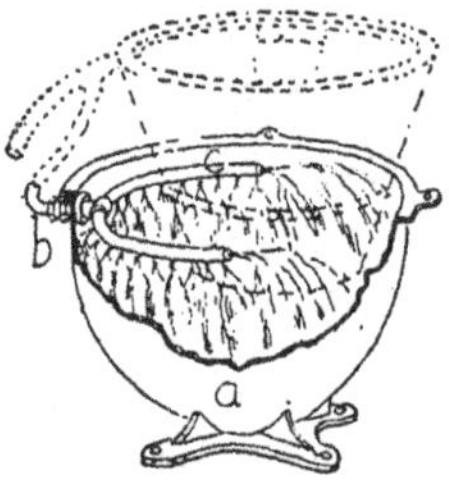

Fig. 118.—Container and Sprinkler.

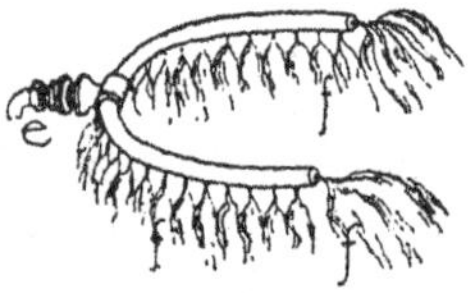

Fig. 119.—Sprinkler.

Tylor's Pan-Closet.

a, Receiver or container. *b* Supply-pipe. *c*, Sprinkler. *e*, Coupling. *f*, Jets of water.

and collect the spatterings from the container. A pan-closet with an earthenware container that has a flushing arrangement and is properly ventilated, the pan being opened or shut by a simple crank and held in position by a weighted lever, would probably form one of the best arrangements for a closet of this class. Underhay, of London, also supplies a white earthenware container with the pan-closets which he manufactures, when there is a call for them. It seems remarkable, when their faulty construction is considered, that such a large number and variety of these closets should have been used in this country.

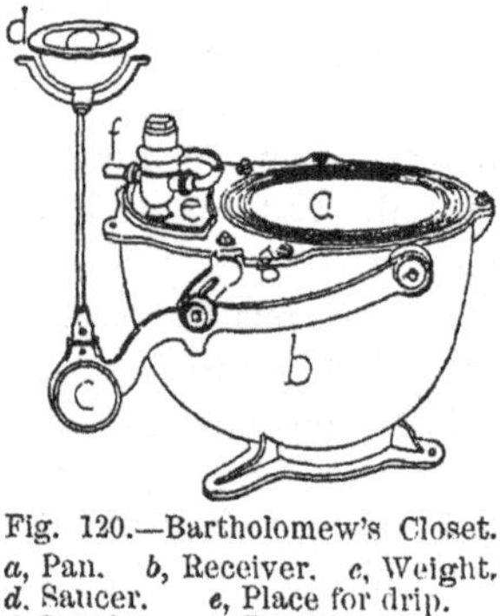

Fig. 120.—Bartholomew's Closet.
a, Pan. *b*, Receiver. *c*, Weight. *d*. Saucer. *e*, Place for drip. *f*, Supply. *g*, Top of receiver.

Bartholomew Pan-Closet. — The Bartholomew closet takes its name from the supply-valve, which was patented in 1854–1858. This closet, if the number used were a correct criterion, might be considered one of the best of the class. I give the illustration of this closet as it is manufactured by Henry

Huber & Co., of New York; it has a simple weighted lever, pan, container and bowl. The supply-valve, on which the novelty is claimed, is screwed to the top of the container, and has a dish or basin formed by a rim raised on the top of the container, to catch any leakage from the supply-valve.

The "Monitor" Closet.—The same manufacturers supply a closet patented by W. S. Carr in 1872, under the name of the "Monitor" closet, which has a piece bolted to the side of the container, that can be easily removed. This is intended to be used in case the pan should need repair, or the container need cleaning out, contingencies which may be looked upon as certainties. "All the trouble of shutting off the water, taking down the seat and detaching the bowl from the top of the receiver is avoided." In the last-mentioned closets the bowl is simply set on the receiver, the bottom projecting through the hole made for the purpose. There is no arrangement for screwing or clamping to the receiver. The "O. I. C." closet is the "Monitor" closet without the removable plate. In the illustration of the "Monitor" closet, one of Carr's supply-valves is shown in position. Huber & Co. Supply these closets with the receivers tarred, zinc-coated (galvanized), or enamelled, and they may be attached to a tank or cistern, instead of directly to a water-main, for the supply to flush them. Vent-couplings are also furnished, when desired, that can be connected with the top of the receiver by

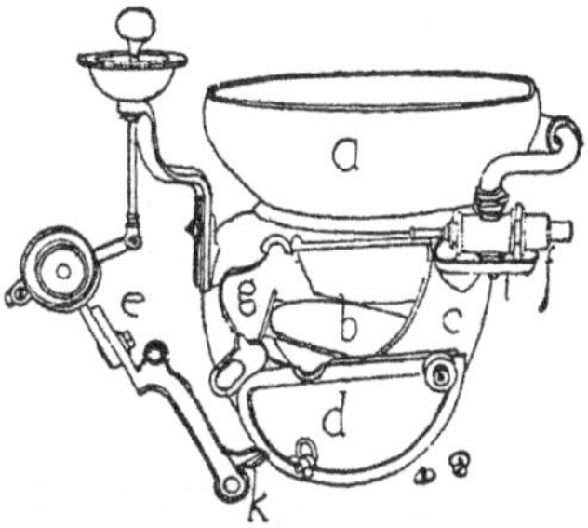

Fig. 121.—"Monitor" Closet with side-piece removed.

a Bowl. *b*, Pan. *c*, Container. *d*, Removable piece. *e*. Adjustable weighted lever. *f*, Supply-pipe and valve. *g*, Spindle. *k*, Stud for operating supply-valve.

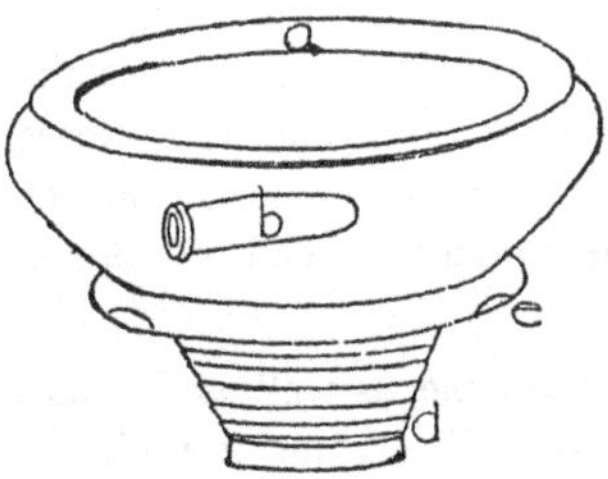

Fig. 122.—Circular Bowl.

means of threads cut into the top of the receiver, into which the brass coupling may be screwed, and a lead pipe soldered on it. The lever in the " Monitor " closet is made so it can be adjusted to different lengths.

Harrison's Pan-Closets.—Chas. Harrison, of Philadelphia, manufactures a number of pan-closets that differ from each other in the position or mechanism of the supply-valve, or in the weighted lever. Harrison furnishes with his pan-closets, when desired, a pan made of rubber, of which he says: they "are non-corrosive, and are impervious to acids." Bowls are manufactured for pan-closets

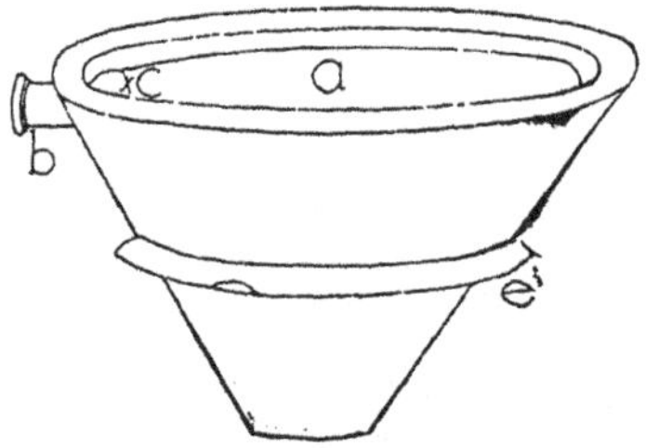

Fig. 123.
Oval Bowl.

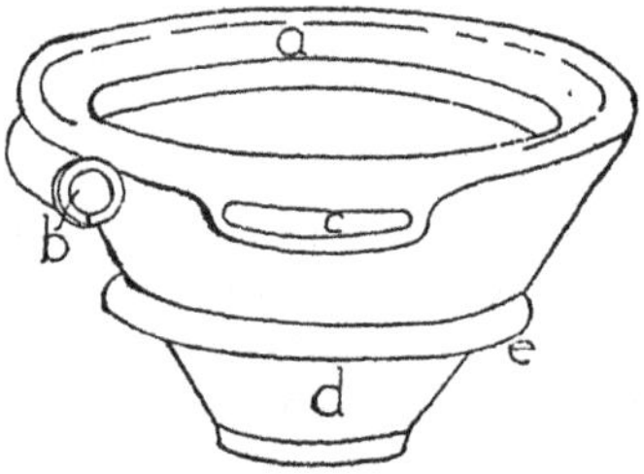

Fig. 124.
Oval Bowl with Vent and Flushing-rim.

a, Flushing-rim. *b*, Orifice for supply-pipe. *c*, Vent. *e*, Rim to rest on top of container. *d*, Portion of dip into pan.

in a variety of shapes, and it seems appropriate to give a description of the different forms in connection with this class of closet, although the same style of bowls are sometimes used in connection with iron hoppers, and properly come under the head of hopper-closets.

Pan-Closet Bowls.—The most objectionable, as well as the most common form of closet bowl is circular. The water to flush the bowl comes through an opening a little less than half an inch in diameter, when the lead pipe has been inserted into the earthenware orifice made for the purpose. The water when turned on forms a puny, spiral stream, that does little if any good in flowing around the sides of the bowl. These bowls are sometimes furnished with flushing-rims. There are conical bowls in which the supply of water is brought in at right angles with the circumference,

Fig. 125.—Old

Fig. 126.—New.

Ship Closet-Bowls.

the water being spread over the surface by means of a fan of metal screwed on the bowl. The orifice in which the lead pipe is inserted is generally an inch and a quarter in diameter in this form of bowl. These bowls are manufactured with an opening near the top, in which a pipe to ventilate the bowl of the closet may be inserted.

There are two kinds of bowl manufactured for ship-closets, both of which are sometimes used in connection with valve-closets. One is conical, the other is hemispherical. The Enterprise Pottery Company make what they call a square French closet-bowl. This bowl has a flushing-rim, and answers for a drip-tray safe and a slop-tray, as well as for a urinal.

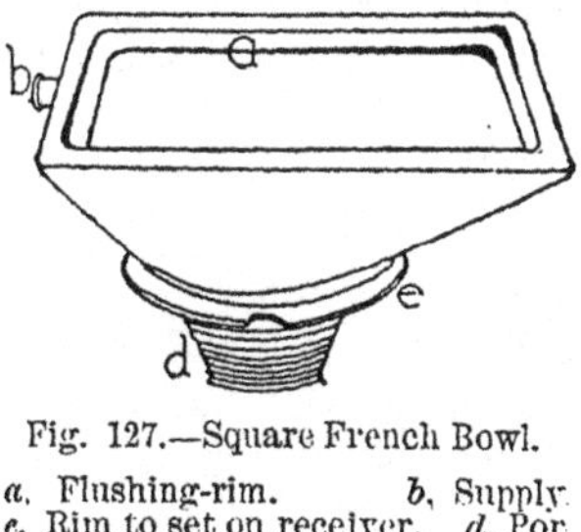

Fig. 127.—Square French Bowl.
a, Flushing-rim. *b*, Supply. *e*, Rim to set on receiver. *d*, Portion that dips into pan.

This form of bowl is used by A. G. Myers, of New York, on his pan-closets. In 1876, Harrison, of Philadelphia, received patents on the combination of a closet-bowl and a drip or slop tray in one piece of earthenware. The bowls last mentioned have forms convenient and useful for the purposes they are intended to fulfil. Harrison's bowl, having the smallest surface exposed to fæcal matter dropping upon it should be preferred.

Among other pan closets, I will mention Jones's "Em

pire" closet, Harrison's "Eureka" and "Ne Plus Ultra" closets, Craigie's "Century" closet, which differ from the ones already described only in the form and position of the weighted lever or supply-valve. The "Cistern" pan-closet, as its name implies, is arranged without a supply valve, being flushed by water from a tank or cistern.

In addition to the nuisance created by the accumulation of filth in the container, the supply-valves, as they are usually made and attached to pan-closets, will generally be found out of order and in a leaking condition, and the joint between the bowl and container is rarely in a perfect state.

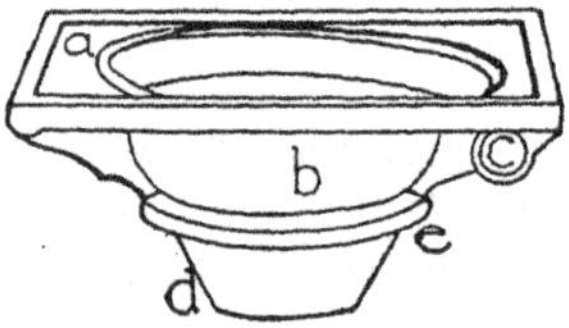

Fig. 128.—Harrison's Bowl.
a, Drip-tray. *b* Bowl. *c* Supply. *e*, Rim to rest on top of container. *d*, Part to form water-seal in pan.

A container with a vent-pipe and flushing arrangements, and made of earthenware or enamelled iron; a bowl with a flushing-rim and at least an inch-and-a-quarter inlet for the supply-pipe, and the water-supply taken from a small tank or cistern placed directly over the closet, should be obtained, if at any time it be absolutely necessary to select a closet of this class. When the improvements above enumerated have been applied to a pan-closet, it loses its only hold on the people—cheapness.

SECTION IV.

PLUNGER-CLOSETS.

THERE is evidence that closets which properly belong to to this class were in use both in France and England more than one hundred years ago (See Figures 19–25); in fact, this appears to be the only form or class of modern closet which it can be positively asserted was in use at this early date.

It is a strange fact that, while such a large number of patents were issued for improvements or variations on valve and pan closets, not a single attempt was made to improve on the plunger-closets, as no patent was granted until the year 1857, when a patent was issued by the United States for a tank connected with a solid plunger-closet, invented by Henry & Campbells. Under this class properly come all closets that have a plug instead of a valve or pan to break the connection between the soil-pipe or trap and the bowl. The plug may be solid or hollow, and have its seat over the trap or the soil-pipe or over an offset from the soil-pipe. The plug is raised and lowered into position by a

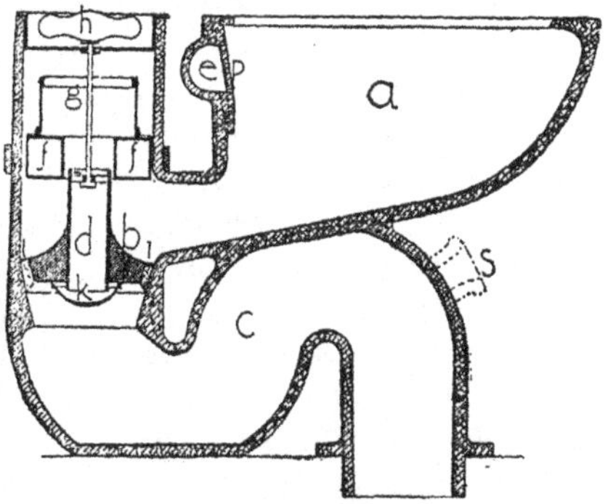

Fig. 129.—Section.
Jennings & Lovegroove's Plunger-Closet.
a, Bowl. *b*, Plunger or plug. *c*, Trap. *d*, Overflow through plunger. *e*, Supply-orifice. *f*, Float for governing the supply-valve. *g*, Lever connecting float with supply-valve. *h*, Hand-pull. *i*, India-rubber band around plunger. *k*, Saucer-trap to the overflow. *l*, Support for supply-valve. *m*, Cross-bar in plunger. *p*, Fan. *s*, Dotted line to show where vent-pipe is formed on Jennings's closet at the present time. *o*, Supply-valve.

short rod, which is attached directly to the plug, and at the top of which is the knob or handle for the hand. The bowl is usually kept about half full of water by the plunger. An overflow is provided through the plunger, when it is hollow, and through a separate pipe usually having a water-seal, when it is solid.

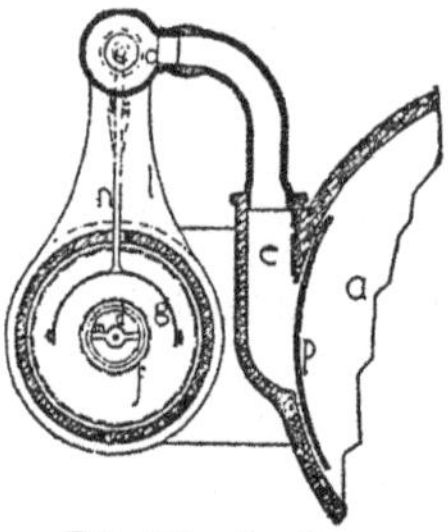

Fig. 130.—Section. Showing Plunger-Compartment and Supply.

Jennings & Lovegroove's Plunger-Closet. The first letters patent for a closet of this class issued by Great Britain were issued to George Jennings & Lovegroove, in the year 1858. This might, with propriety, be named as the first invention in the class, containing as it does a novelty in the hollow plunger; the solid plunger being old. Since the date mentioned, Jennings has received a number of patents for alterations in the form and position of the different portions of his closet, which he considers improvements on his first invention. Jennings appears at first to have received the unqualified support of, and recommendations from sanitary engineers and architects; but those who praised his closet most highly have seen reason to withdraw or at least qualify their recommendations. The chamber which contains the float for closing the supply-valve, and also the chamber in which the plunger moves, were connected with the bowl of the closet, and it has been found in practice that this chamber, plunger and float become foul by sediment composed of excremetitious matter collecting on them, as well

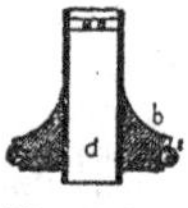

Fig. 131.

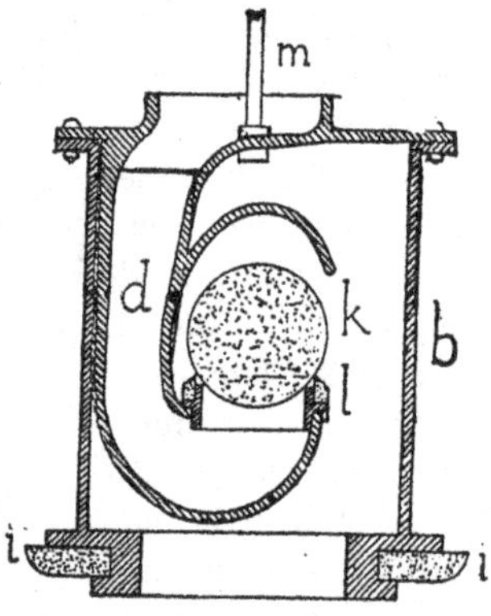

Fig. 132.—Two Forms of Plunger.
b, Plunger. *d*, Overflow. *i*, India-rubber ring. *k*, Ball-valve of rubber. *l*, Rubber seat with knife-edge.

as larger particles lodging in the concealed parts of the closet, in their passage to the trap. The overflow and plunger have a number of different forms. As first invented (Fig. 129), the closet was made in one piece of earthenware. The supply-valve is operated by a float, which encircles the top of the plunger. The mouth of the siphon trap is in the shape of an inverted frustum of a hollow cone, and into this opening the plunger, encircled by a rubber band, fits tightly. The central part of the plunger, which is hollow, and forms the overflow, dips below the water in a small saucer, in this manner forming a water-seal trap. The saucer is suspended from the plunger by means of small hooks and eyes. Jennings on some occasions uses a plunger in which the overflow has a mechanical valve similar to the ball-valve described in connection with the overflow of his valve-closet. This is used in connection with his trapless closet. The ball in this valve is composed of india-rubber, and has its seat on a rubber band or ring, which has a knife edge. On this seat the ball will adjust itself, a slight inequality in the ball, or a small piece of foreign matter making little difference in the efficiency of the joint.

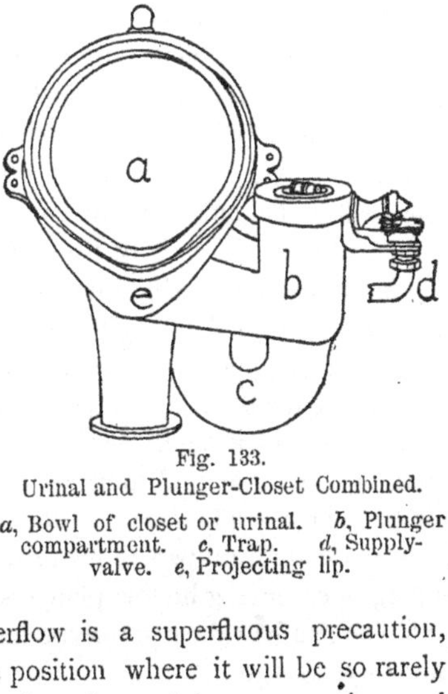

Fig. 133.
Urinal and Plunger-Closet Combined.
a, Bowl of closet or urinal. *b*, Plunger compartment. *c*, Trap. *d*, Supply-valve. *e*, Projecting lip.

A mechanical trap in the overflow is a superfluous precaution, unless the closet is placed in a position where it will be so rarely used that the water-trap might lose its seal by evaporation and thus become inoperative. In a closet like Jennings's, the supply-valve being governed by a float, the bowl would be kept full of water, provided the supply-valve did not get out of order; but the siphon trap under the closet, if the water should evaporate, would allow the gases to get into the room through the plunger-chamber.

These closets are so arranged that they may be supplied either from a tank or cistern, or directly from the supply-pipe through one of Jennings's hydrostatic diaphragm supply-valves. This valve is so delicate, and requires such nice adjustment, that it is rarely in proper working order. A. E. Jennings, of New York, furnishes these closets with or without a trap in one piece of earthenware, or with an earthenware bowl, a plunger-compartment, and trap or offset of iron. The trap is properly provided with a vent-pipe at the crown, on the side connecting with the soil-pipe.

The Jennings "Night Urinal and Water-Closet" is a fixture, with a high earthenware back and projecting lip, intended to stand without wood-work, made in one piece of earthenware, for bowl, plug, chamber and trap.

For the purpose which Jennings claims utility in his closet, and under such circumstances if they should arise, a closet of this class would undoubtedly be useful. He says: "the object being to save water from sinks, waste-pipes, or other sources where the water-supply will be limited." Jennings's closets have been used extensively in all civilized parts of Europe and America.

Quite a number of English and American inventors have followed in Henry & Campbell's, and Jenning's and Lovegroove's steps, making supposed improvements on their closets, the only real improvements being in the form of the supply-valve and in contrivances for flushing the plunger-chamber. The variations consist in the form and position of the supply-valves, plunger-chamber, plugs or plungers, and the overflow. Closets of this class divide naturally into two forms or types, those which follow the Jennings idea having hollow plungers, while the closets which have solid plungers had their prototype in the closets used more than a hundred years ago.

Renison's Plunger.—The second patent for a hollow plunger was issued to Thomas Renison, a plumber, of Glasgow, by Great Britain, in 1862. The plunger is so arranged that the pressure of compressed air confined within the plunger will retain the water at any desired height in the bowl; but when the water rises above the height for which the mechanism is arranged, it overcomes the force of the air and flows through the plunger. This plunger is formed by a small cylinder placed within a large one, a space for air being left between them, and the top, to which the hand-pull is attached, is covered over without an opening. The plunger fits into a similar double compartment, which forms a part of the plunger-chamber. This closet is not practicable, as the chamber in which the plunger fits would in a little while be filled with excrementitious matter, so the plunger would be useless.

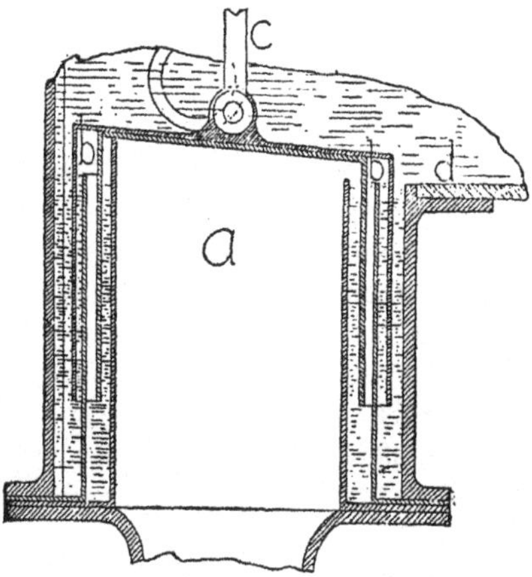

Fig. 134.—Renison's Plunger.
a, Plunger. *b*, Compressed air. *c*, Hand-pull. *d*, Bowl.

Baldwin Latham's Closet.—Baldwin Latham, in his work on sewerage, describes a plunger-closet which he designed as an improvement on the simple hopper-closet. The bowl and plunger-compartment are in one piece. The plunger, which is hollow, retains a large amount of water in the bowl. As is the case in other closets of this class, when the plunger is raised a large body of water is discharged through the soil-pipe, in this manner tending more effectually to scour it than would otherwise be the case. The bottom or outlet from the bowl, and the opening into the soil-pipe are so arranged as to form a water-seal trap. The plunger-compartment has a special vent-pipe, and the bowl a flushing-rim, while the rod connected with the hand-pull passes through

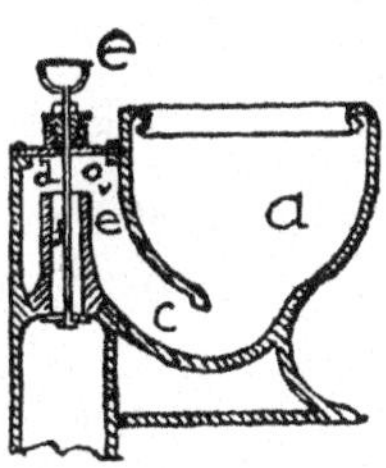

Fig. 135.
Latham's Closet.
a Bowl. *c*. Trap.
d, Plunger-chamber.
b Vent. *f* Overflow.

a stuffing-box, thus effectually preventing the passage of gas at this point.

Tylor & Sons' Plunger Closet.—Tylor & Sons, of London, in 1873 invented a plunger-closet, the novelty of which consisted in the form of the plunger and trap under the closet. The plunger is hollow, the hand-pull being attached to it by means of a bar which extends across the plunger on the inside. The bottom of the plunger is bell-shaped, with a ridge encircling it. Over this part of the plunger a rubber cylinder or pouch is stretched. This plunger has its seat on a rubber band. In this manner it is pro-proposed to make a tight joint. The plunger forms the overflow, and has its seat on the opening of a U-shaped trap; the other end opens into a chamber

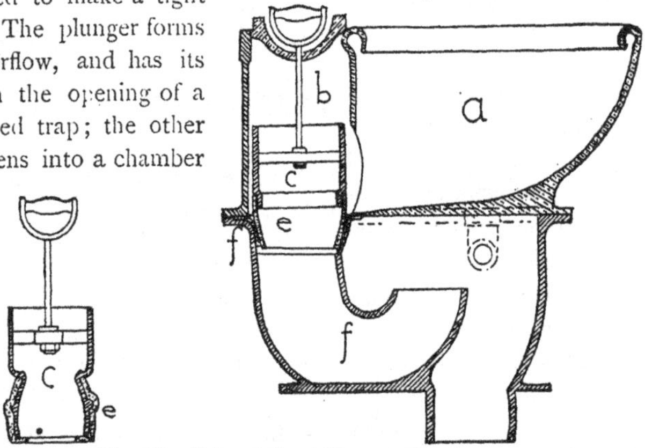

Fig. 136.—Tylor & Sons' Plunger-Closet.
a, Bowl. *b*, Plunger-chamber. *c*, Plunger. *f*, Trap. *e*, Rubber band.

which is joined to the soil-pipe. There are a number of corners and hollows in a trap of this kind, that would be sure to collect and retain filth, and for that reason it should be avoided.

Demarest's Plunger-Closets.—J. L. Mott & Co. manufacture several forms of plunger-closets: the "Demarest," the "Hygeia," and "Premier." Each has a bowl with a side outlet, while the earliest pattern has a compartment that contains the plunger, a float for governing the supply-valve, and the supply-valve. The plunger is a hollow cylinder closed at the top. The overflow is through an opening in the side, from below which a partition rises to a slight distance above the top of the opening. In this manner a water-seal trap is formed to the overflow. The bottom of the plunger is encircled by a rubber band or ring, which has a seat on the top of the soil-pipe offset or trap. The float is connected with the supply-valve by means of a lever that encircles the rod of the hand-pull, the float and supply-valve being on opposite sides of the plunger, so that the flow of water would be shut off when it had reached the top of the overflow. The upper part of the supply-valve has a row of small holes around it, so arranged as to throw small jets of water into the plunger-chamber. These jets are for the purpose of washing off any sediment or waste matter that may collect on the plunger, float, or the sides of the

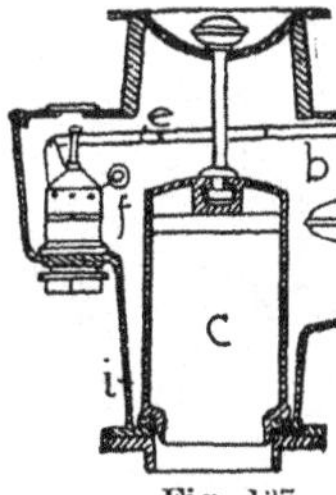

Fig. 137.

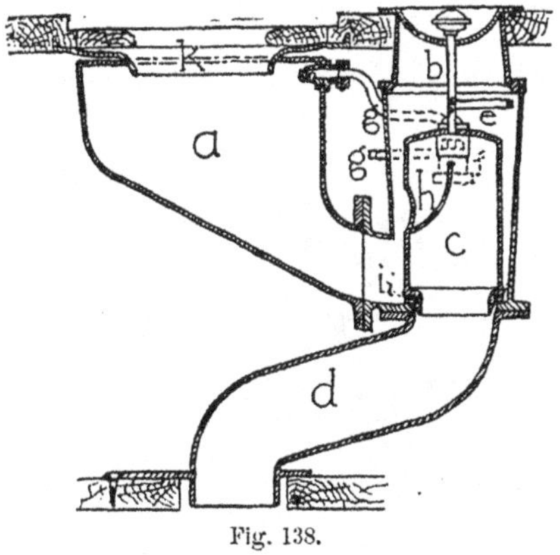

Fig. 138.

Demarest's Plunger-Closet.

a, Bowl. *b*, Plunger-chamber. *c*, Plunger. *d*, Offset. A siphon-trap may be put in same place. *e*, Lever connecting float and supply-valve. *f* Supply-valve. *g*, Supply-pipe *h*, Overflow. *i*, Rubber band around the plunger. *k*. Slop-safe or urinal. *l*, Float. *o*, Perforations in supply-valve for washing plunger-compartment.

compartment. This closet has been extensively used in this country, and has been patented both in this country and in England. John Demarest designed the "Hygeia" and "Premier" as an improved form of plunger. The plunger consists of a simple cylinder, open at each end, with a neck formed on the bottom, around which a vulcanized rubber band, circular in section, is stretched. This band enables the plunger to make a tight joint on its seat. The overflow is between the plunger-chamber and the plunger, and forms an opening about an eighth of an inch around the plunger. The hand-pull has a short lever bolted to it, the other end of which is connected by a chain to a tank or cistern, from which the water is conducted by an inch-and-a-quarter pipe to a flushing-rim around the closet; in this manner the sides of the closet are washed more efficiently than they would be with a simple pan.

The "Premier" closet differs from the "Hygeia" only in the fact that it is arranged for a supply-valve instead of for a cistern. The valve is attached to the top of the plunger-chamber, and is operated by a short lever similar to the one used in connection with the tank. These closets have porcelain bowls and enamelled iron plunger-

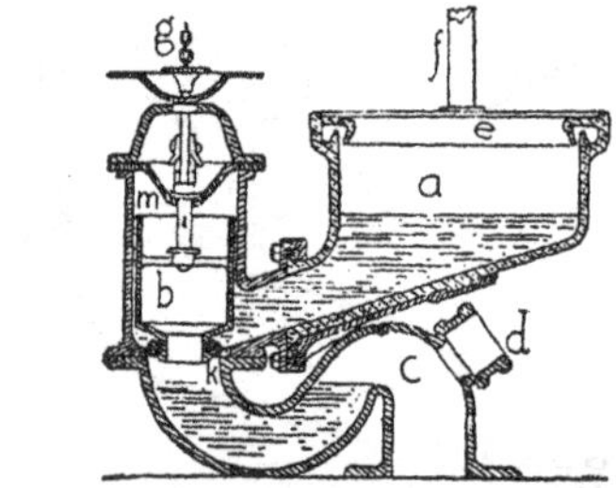

Fig. 139.—Section.

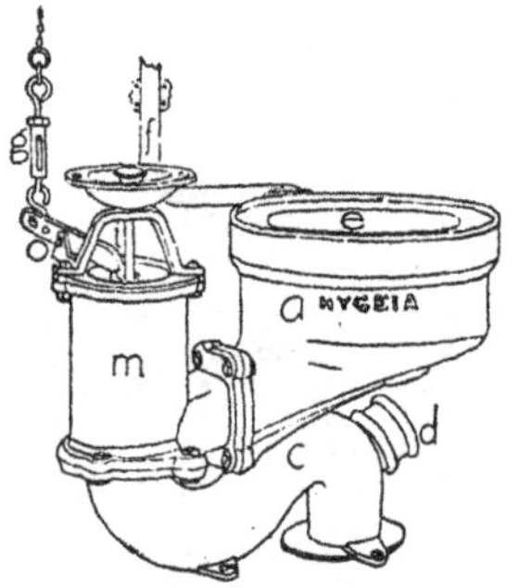

Fig. 140.—Perspective.

Demarest's Hygeia.

a, Bowl. *b*, Plunger. *c* Trap. *d* Vent to trap. *e*, Flushing-rim.
f, Supply from cistern. *g* Chain to cistern. *i*, Hand-pull.
k, Rubber ring. *m*, Plunger-chamber. *o*, Lever.

chambers, while the plunger and the cover for the chamber, bridge and lever are zinc-coated (galvanized), and put together by means of screw-bolts and clamps. They have above the floor a straight outlet, an offset, or a trap. The trap and offset have an opening at their crown, with a bell into which a vent-pipe must be caulked. Where the trap or outlet passes through the floor there is a broad flange, which is arranged so as to be screwed to the floor over the soil-pipe.

For a plunger-closet, these closets are probably arranged as well as a closet of this class could be made, as they have a small plunger-chamber, no float, a rubber ring on the bottom of the plunger, a flushing-rim to the bowl, a siphon vented trap above the floor; while the different parts are thoroughly put together, well made and of good materials. The only arrangement devised for washing the plunger-chamber consists in the water passing between the plunger and its chamber to the overflow.

The Hartford Glass Closet.—The glass closet which has been recently introduced into the market was invented in this country by M. Hogan in 1878, and he also received patents for improvements in the year 1882. In addition to the novelty in the form of the overflow, which is intended to take place through an opening in the plunger, this closet is provided with a lining or

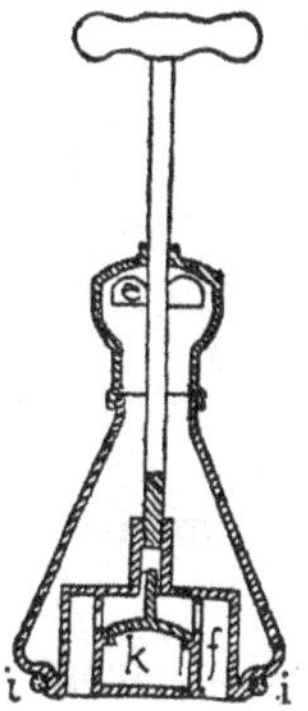

Fig. 141.

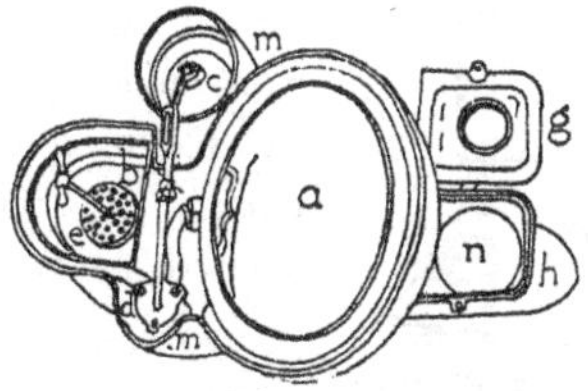

Fig. 142.

Hartford Glass Closet.

a, Bowl. *b*, Plunger-chamber. *c*, Float. *d*, Valve for supply. *e*, Top of plunger, showing perforations for overflow. *m* Iron casing. *g*, Inspection opening. *h*, Trap. *n*, Flap-valve. *f*, Overflow. *k*, Mechanical valve. *i*, India-rubber ring.

inner shell composed of glass, on the supposition that glass will not have the small net-work of cracks which cover the glazed surface of most earthenware. These small cracks, the "craze," in earthenware serve as a nucleus around which filth and sediment collect. Of course glass can be used as readily in this connection with a valve or hopper closet as with the one under consideration, and to their advantage, if the claim made of the non-collection of sediment can be sustained. The plunger in this closet has an overflow through perforations on top of the plug into a water-seal trap situated in the bottom of the plunger. The overflow also has a mechanical valve as an additional seal against sewer-air. The valve is so arranged that the water flowing from the bowl of the closet would raise it from its seat, while it would drop into its place again as soon as the pressure of the water should be removed. The water-seal trap is arranged either as shown in the illustration or in the form of a simple U-trap, with the mechanical valve on the top of the short arm of the trap. The siphon trap under the closet has a flap-valve and inspection-cover bolted on near its crown, the flap-valve being intended to form a fourth intervention between the soil-pipe and closet-bowl.

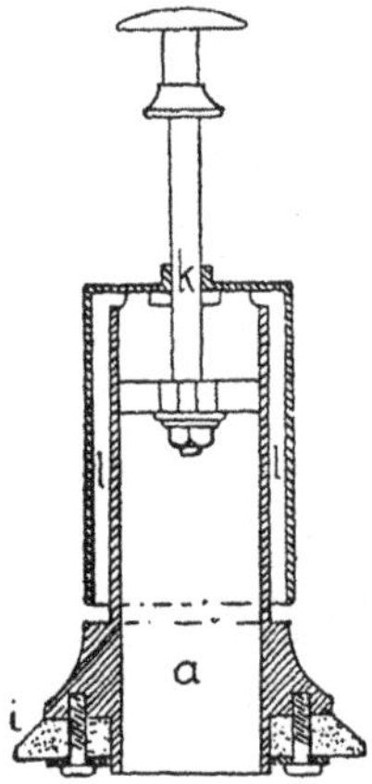

Fig. 143.—Blunt's Plunger.
a, Small cylinder.
l, Space between cylinders.
i. India-rubber flange.
k, Hand-pull.

Among other closets belonging properly to this class and type, in which the plunger is dfferent from the ones previously described, I find one invented by Blunt in 1876, one by W. Turton, 1878, one each by Keith and Blesch in 1882. These closets all received patents from the United States.

The Blunt Plunger.—The Blunt plunger consists of one cylinder, closed at the top, slipped over one of a smaller diameter, which is open at both ends. When the two cylinders are joined together there is a passageway for the water through the space formed by the difference in their diameters, the outside cylinder being closed at the top and open at the bottom, while the inside cylinder is open both at the top and the bottom. The water which stands in the space between the two cylinders forms a water-seal equal to the height of the inside cylinder. This plunger has a rubber flange fastened to the bottom of the plunger by screw-bolts.

The Keith Plunger.—Keith's plunger consists of a cup or bell fitted into a receptacle made for the purpose on the top of the plunger. The water-seal trap is formed on the same principle as that of the bell-trap in such common use for area cesspools. The seal in the trap is slight and would be easily destroyed by evaporation.

The Blesch Plunger.—Blesch's plunger is arranged with a supply-pipe running down into the centre of it. In this manner the centre of the plunger, which is hollow, is filled with water when the

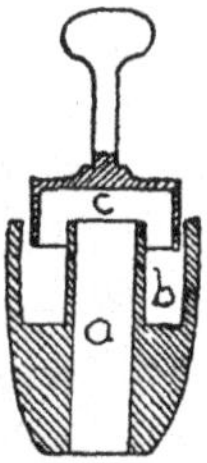

Fig. 144.
Keith's Plunger.

a, Overflow.
b, Trap.
c, Bell.

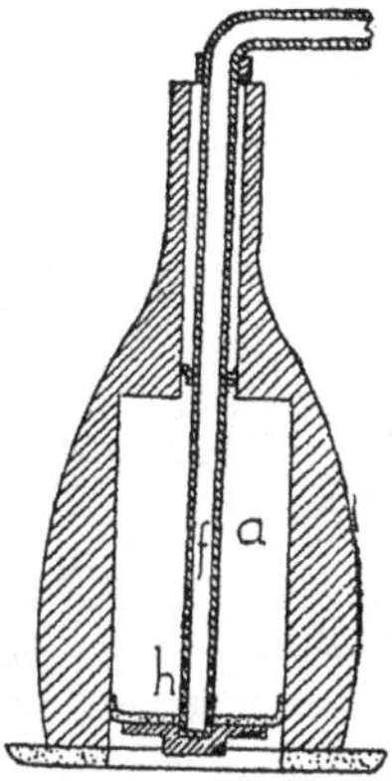

Fig. 145.—Blesch's Plunger.
a, Hollow in centre of plunger.
f, Supply-pipe.
h, Piston on supply-pipe.
i, India-rubber bottom.

plunger is lifted from its seat. The water can only escape through small holes left for the purpose, and consequently the plunger takes its seat gradually. The illustration shows the plunger raised to its highest point.

SOLID-PLUNGER CLOSETS.

Closets of this type cannot claim a novelty on the distinctive feature of the class, as the solid plunger is old, unless it be for the shape, position, or manner of lifting. The first closet of this type and class, and also the first plunger-closet for which a patent was issued, was invented in this country by Henry & Campbells in 1857, being one year earlier than the invention of Jennings in England.

Henry & Campbells's Closet.—There was nothing new in either the tank, which was placed under the seat, or in the plunger, which was solid, when this closet was patented. This closet, as will be seen by the illustration, contains a large amount of water for scouring the soil-pipe. When the plunger is lifted from its seat, the contents of the tank, plunger-chamber and bowl would be emptied into the soil-pipe. The interest in this closet consists in its being the first of the class patented.

Fig. 146.—Section.

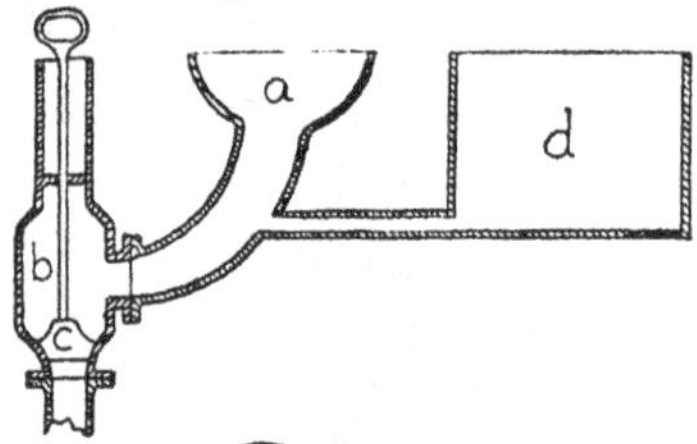

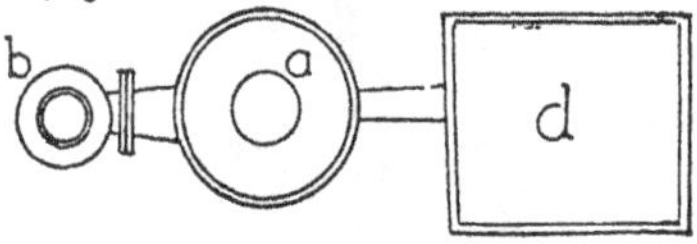

Fig. 147.—Top View.
Henry & Campbells's Closet.
a, Bowl. *b*, Plunger-chamber. *c*, Plunger. *d*, Tank.

Conron's Closet.—The United States in 1868 granted patents to G. Conron for a solid plunger-closet. In this closet the bowl is conical, while the outlet is at the bottom; thence it passes through a horizontal pipe about eighteen inches long, at the end of which is situated the plunger-chamber. Between this chamber and the bowl is a branch pipe, which serves as a conduit for water into the float-chamber that contains the float governing the supply-valve. The overflow opens near the top of the plunger-chamber, and connects with the soil-pipe below the seat of the plunger, this

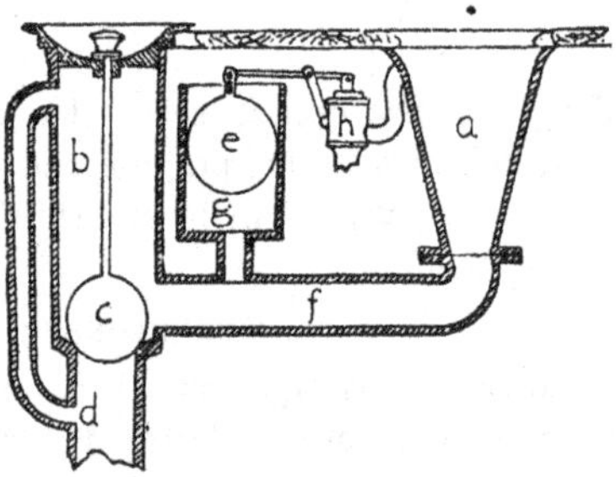

Fig. 148.—Conron's Closet.
a, Bowl. *b*, Plunger-chamber. *c*, Plunger. *d*, Overflow. *e*, Float for supply-valve. *f*, Horizontal connecting-pipe. *g*, Float-chamber. *h*, Supply-valve.

part of the closet being a solid sphere joined directly to the hand-pull by a simple rod. J. J. Frey improved this closet by adding a flushing-rim.

Ingleton's Closet.—In 1869 a closet belonging to this type was invented in England by James Ingleton. This closet is a good example of the form with a solid plunger. The plunger and float compartments are separate from the bowl and each other, being

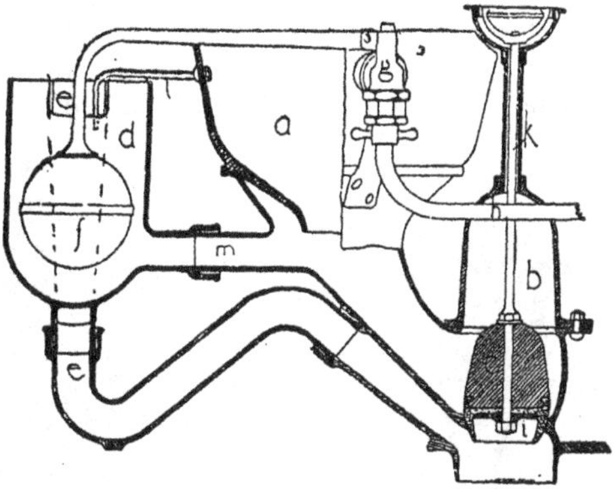

Fig. 149.—Ingleton's Closet.

a, Bow. *b*, Plunger-chamber. *c*, Cast-iron block on bottom of plunger. *i*, India-rubber bottom. *d*, Float-chamber. *e*, Overflow. *f*, Float. *g*, Supply-valve. *h*, Supply-pipe. *k*, Rod for hand-pull. *l*, Driblet to fill trap of overflow. *m*, Connection between bowl and float-chamber.

only connected by short pieces of pipe. What is gained by having these compartments only large enough for the plunger and float to move in them is more than counterbalanced by the disadvantage of the pieces of pipe connecting the bowl with the two chambers. Such are always to be avoided on account of the difficulty in keeping them clean. The overflow is from the top of the float-chamber through a siphon-trap, and empties into the soil-pipe just below the seat of the plunger. The overflow is kept full of water by a driblet. The top part of the plug of the plunger is made of cast-iron, being intended to act as a weight and keep the plunger in place, while the bottom is a slightly conical piece of vulcanized india-rubber, the part that fits into the seat being in the form of a projecting lip, so as to be flexible.

Pearson's "Twin-Basin" Closet.—The "Twin-Basin Closet," invented in England by Pearson, in 1874, appears to have been extensively used in that country, and also to have had a limited sale in this country. This closet was divided into two compartments of almost equal dimensions, one for the bowl, the other for the ball-cock-float and plunger. This closet when first introduced was without an overflow, but the manufacturers found it necessary to supply the closet with an overflow which is taken from the top of the plunger-chamber and is sealed by water before it enters the

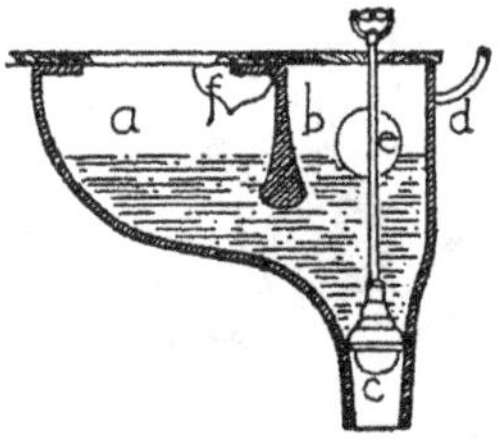

Fig. 150.—Section of Closet.

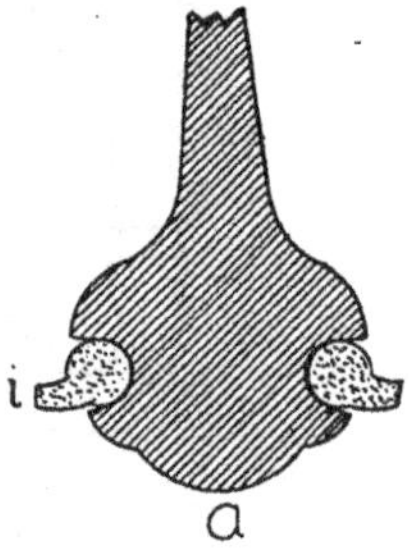

Fig. 151.—Section of Plunger.

Pearson's Closet.

a, Bowl. *b*, Plunger-chamber. *c*, Plunger. *d*, Supply-pipe. *e*, Float for ball-cock. *i*, Rubber band around plunger.

soil-pipe below the plunger. The ball-cock is not as delicate as the hydrostatic diaphragm valve; it does not require the nice adjustment, and is less liable to get out of order.

Zane's Closet.—The "Sanitary" Closet is manufactured by Joseph Zane & Co., of Boston. The bowl, of porcelain, is attached to the plunger-chamber, which is of zinc-coated (galvanized) or enamelled iron, by means of clamp-screws. The supply-valve is a ball-cock designed specially for this closet. The plunger is solid and has a rubber ring around the bottom. The overflow is through a balanced pan or saucer that will tilt when it has received a certain amount of water, thus allowing any excess of water to pass through it. There is a small driblet from the supply-valve, placed so as to drip into the pan of the overflow. The plunger-chamber and the bowl connecting with the plunger-chamber have large openings to which vent-pipes, which are inserted into a heated flue are to be connected. These vent-pipes are intended to cause a current of air, created by the heat in the flue, to pass from the room through the two compartments of the closet, thence into and up the flue, taking with it any offensive or injurious gases, left by the actual passage of fæcal matter, or arising from the sediment that adheres to the sides of the plunger-compartment. This closet takes a large amount of water when used, but the simplicity of the mechanism, the enamelled plunger-chamber and vent-pipes, make it one of the best of its class in the market. This closet should be supplied with a siphon-trap, properly vented.

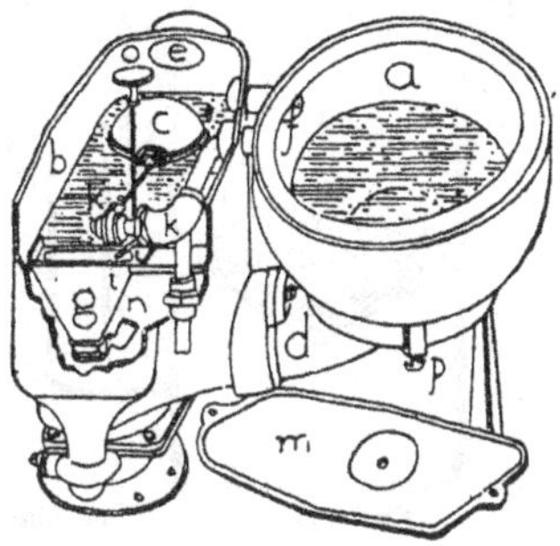

Fig. 152.—Zane's "Sanitary Closet."

a Bowl. *b*, Plunger-chamber. *c*, Float. *d*. Connection between plunger-chamber and bowl. *f*, *e*. Vent. *g*, Overflow. *n*, Balanced pan. *i*, Driblet. *k*. Supply. *m*, Cover. *o*, Hand-pull. *p*, Clamp-screw.

It is furnished with a straight outlet and an offset that is intended to be joined directly to the soil-pipe. It can only receive its supply through a valve, and has no flushing-rim of consequence.

The Gale Closet.—Among other closets manufactured by A. G. Myers, of New York, which belong to this class and type, is one invented by M. F. Gale, in 1880. The peculiarity in this closet consists in the arrangements for the overflow, and for washing the plunger-chamber. The overflow is through an upright pipe, while the float which governs the supply is in the form of a cylindrical cup slightly enlarged, and closed at one end, and it is placed with its open end downward over the overflow pipe. In this manner a water-trap is formed between the plunger-chamber and the soil-pipe. The float is connected with the supply-valve by a crooked lever, while its chamber is separated from the plunger-compartment by means of a perforated partition. The overflow is again trapped by dipping into a small box of water, before it enters the soil-pipe.

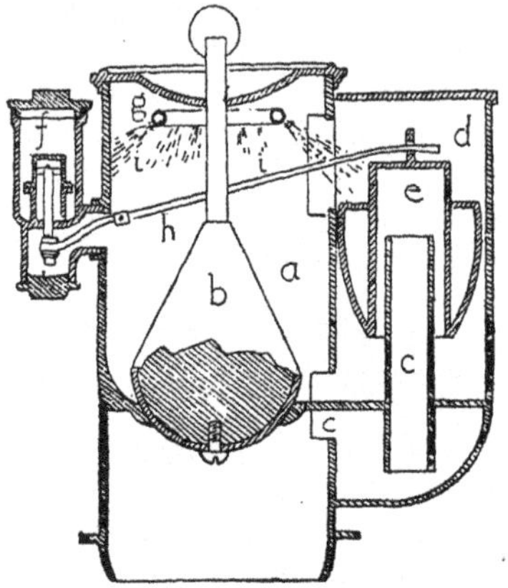

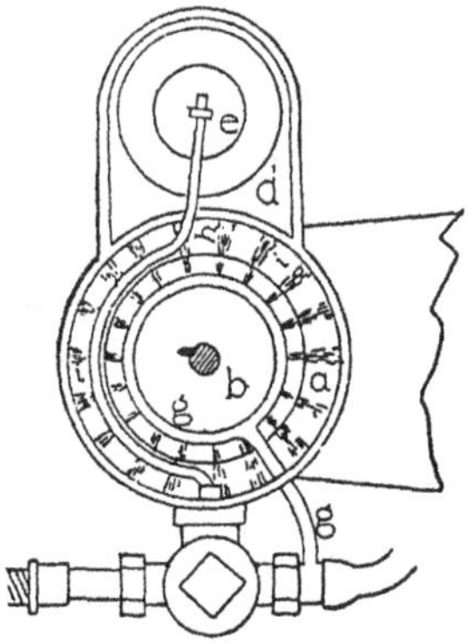

Fig. 153. The Gale Closet Fig. 154.
Section and top view of Plunger-Chamber.

a, Plunger-chamber. *b*, Plunger. *c*, Overflow. *d*, Float-chamber. *e*, Float governing supply-valve. *f*, Supply-valve. *g*, Sprinkler for washing plunger-chamber. *h*, Lever connecting float and supply-valve. *i*, Small jets of water.

The plunger-chamber has a small pipe connected with the supply-pipe encircling it, around the circumference of which are a row of small holes, through which small jets of water are thrown over different parts of the plunger-chamber. In this manner it is proposed to keep this part of the closet clean. The bowl of this closet is porcelain, while the iron parts are all enamelled and are connected with the bowl by means of set-screws.

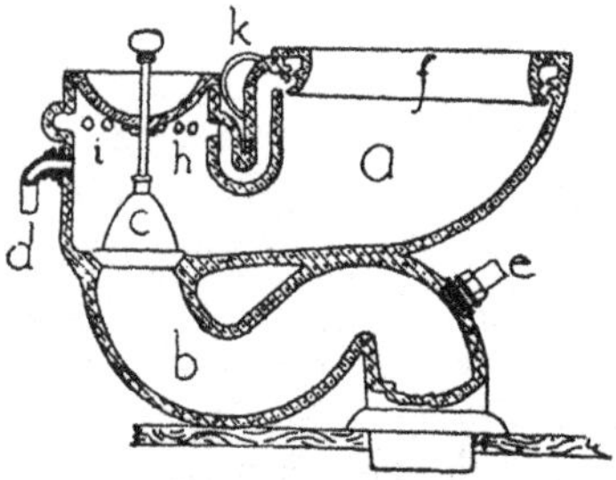

Fig. 155.—Section.
Myers's All China Closet.

***a*, Bowl. *b*, Trap. *c*. Plunger. *d*, Overflow. *e*, Vent-pipe coupling. *f*, Flushing-rim. *h*, Plunger-chamber. *i*, Holes for letting water into plunger-chamber. *k*, Supply-pipe.**

Myers's "All China" Closet. The same firm manufactured a solid plunger-closet invented in

1881. "This closet has an independent overflow, a solid plunger, and a flushing-rim with patent jagged points, causing the water to form a cataract....There is also an encircling chamber" around the plunger-compartment, which would throw jets of water in all directions, crossing each other, probably washing a chamber of this kind as effectually as any contrivance of the kind could. The overflow does not enter the soil-pipe, but is intended either to empty into the safe, or it may be carried through the wall, and empty on the outside of the house. The trap has a brass coupling inserted at its crown, to which a vent-pipe may be soldered. An opening for a vent-pipe should at the least be two inches in diameter. The same closet is supplied with a simple offset instead of a trap.

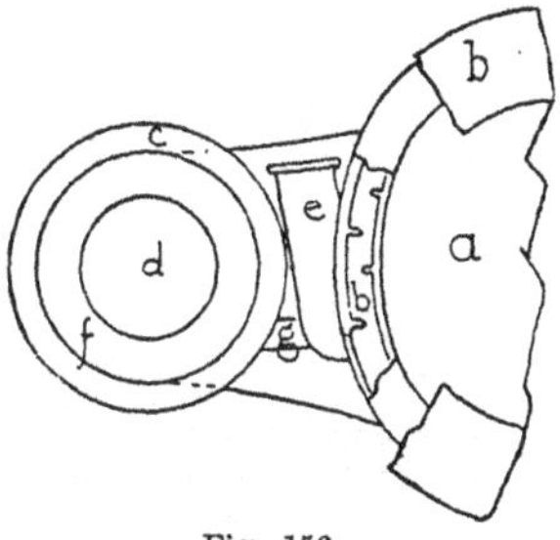

Fig. 156.
Flushing-rim, All China Closet.

a, Bowl. *b*, Flushing-rim on bowl. *c*, Flushing-circle for plunger-chamber. *d*, Plunger. *f*, Plunger compartment. *e*, Supply. *g*, Connection between the two flushing-rims.

Myers also manufactures the "Egg Oval" Closet; the peculiarity of which consists in the shape of the outlet from the bowl. This outlet in cross section is egg-shaped. Such a section would be of value only in large sewers.

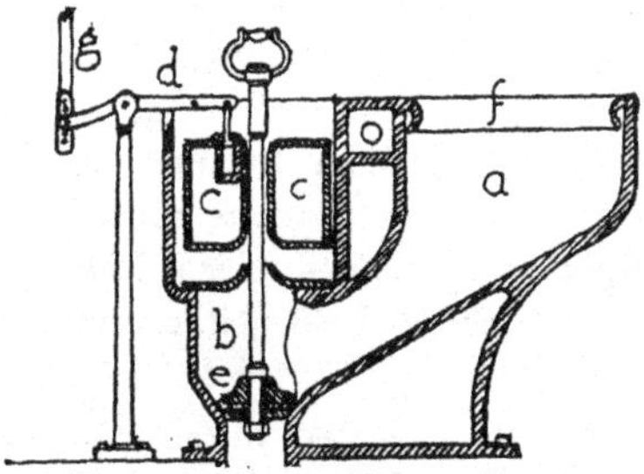

Fig. 157.—Moore's Plunger-Closet.

a, Bowl. *b*, Plunger-chamber. *c*, Float. *d*, Lever. *f*, Flushing-rim. *e*, Plunger. *g*, Rod connecting with tank.

Moore's Closet.—A solid-plunger closet was invented in 1881, by C. H. Moore, of Philadelphia, Penn. This closet is made in one piece of earthenware, and the novelty consists in the float being so arranged that when the water falls to a certain point in the bowl and plunger-chamber, the closet is flushed from the tank. The float, in falling, pulls down the long arm of a lever that is bolted to it. The short arm of this lever is connected with a long upright rod, and the head of this rod projects up into the tank, where it forms a valve which governs the supply to the closet. The necessity for an overflow in this closet is obviated by an enlargement of the arm of the plunger; this enlargement being too great to pass through the float, the plunger would be necessarily lifted from its seat, when the water had risen above a certain height.

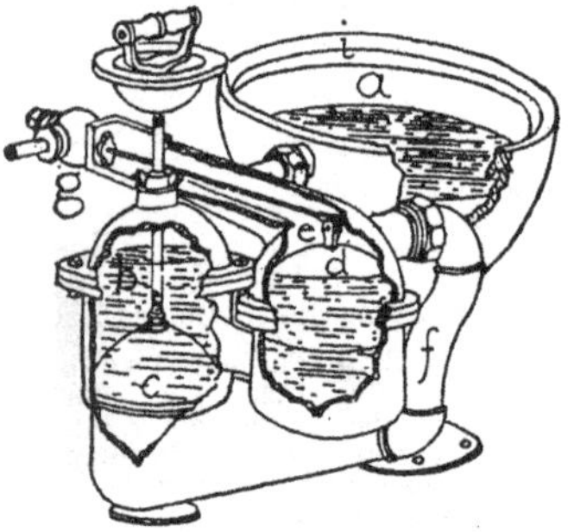

Fig. 158.—Milne and Gant's Closet.
a Bowl. *b*. Plunger-chamber. *c*, Plunger. *e*, Float governing the supply-valve. *g*, Supply-valve. *f*, Overflow. *i*, Flushing-rim.

Milne and Gant's Closet.—In 1880, letters patent were issued to Milne and Gant, of San Francisco, for a solid-plunger closet. This closet has plunger and float chambers, just large enough for the plunger and float to move up and down in them, and they are connected with each other only by small openings. The overflow in this closet rises from the top of the float-chamber, just below the top of the bowl, and enters the offset before it joins the soil-pipe. Although the compartments, as in the Ingleton Closet (Fig. 149), are simply large enough for the purpose they are intended to serve, still the small conduits connecting them would make the scouring and cleaning received from the water passing through it even less effectual than if everything were contained in one large compartment.

Budde's Closet.—A plunger-closet was invented in this country in 1882, by J. Budde. The arrangement of the mechanical valve in the overflow is the new feature of this closet. The overflow is through a tube attached to, and parallel with the plunger-chamber. At the bottom of the overflow there is a ball-valve; this ball is connected by a rod with a hinged float, that covers the top of the plunger-chamber; when the water rises to the top of this chamber, it fills the overflow, raises the valve by means of the float, and passes into the soil-pipe. Neilson, 1873; Gilchrist, 1874; Keyser, 1876; Moellmann, 1882, invented slight and unimportant improvements either in the form and position of the overflow or supply-valve, in connection with which each used a plain solid plunger.

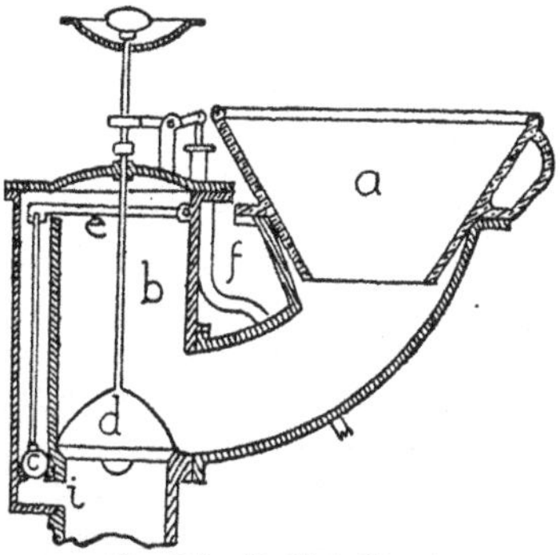

Fig. 159.—Budde's Closet.

a, Bowl. *b*, Plunger-chamber.
c, Ball-valve in overflow. *d*, Plunger.
e, Float for overflow. *f*, Supply-pipe.
h, Supply-valve. *i*, Overflow.

E. S. Hutchinson's Closet has the plug in the same compartment that forms the bowl, in this respect being similar to closets that were in use more than one hundred years ago.

Stanton M. Howard has recently (1884) invented a closet in which the plunger moves in a horizontal position. The plug being conical, it fits tightly into the side outlet of the bowl. The plunger is operated by means of a rack, and ventilated. The pinion is weighted, and by this means the plunger is forced back against its seat.

DEDUCTIONS.

It will have been noticed by those who have read the foregoing descriptions, that the mechanical faults and the sanitary objections to the plunger-closet are almost identical to those mentioned in connection with valve-closets. The utility of the plunger in a closet of this class depends upon its taking a water-tight seat. All the useful plunger-closets have a rubber flange or ring around the bottom of the plunger, and this form of plunger is generally effective, as long as the vulcanized rubber retains its elasticity; when it gets hard the plumber must be called in to put on a new one.

This form of closet is liable at any time to lose the water in the bowl and plunger-chamber, by foreign matter lodging between the plunger and its seat; so it is never safe to have a closet of this kind connected to the soil-pipe without a trap; although the patentees usually claim it as one of the advantages of their closet, that it does not need a trap between it and the soil-pipe. The plunger takes and retains its seat by the action of gravity, and for this reason it is more simple, and a better mechanical device than the valve; requiring no levers, weights, or cranks to operate it, or keep it in position.

It is necessary for a plunger to be heavy, so it can take and keep its seat properly; this weight is often a decided inconvenience and objection where the closet is to be constantly used by delicate women and small children.

Taking a sanitary view of the subject, the greater number of the different patterns of plunger-closets should be considered inferior to the valve-closets, the plunger chamber being—as with a few exceptions it is—larger than the receiver of a valve-closet, and containing the plunger, float, lever, and frequently the supply-valve, upon the different surfaces of which filthy matter is liable to collect and decay.

Mr. E. S. Philbrick says of closets belonging to this class, "The construction of these (supply) valves is various, and their action is not always reliable, which has at times caused considerable annoyance. If these valves can be made more certain in their action, and the float-chamber made more accessible for frequent cleaning, the apparatus can be made a very satisfactory one; but it needs careful treatment and good care."

The supply-valves not being reliable, closets of this class must have overflows.

Where the plunger is hollow, the overflow is carried off through the plunger, but where the plunger is solid, a separate conduit has to be provided: and where this is trapped by a water-seal the water is liable to be drawn from the trap by siphon action whenever the closet is used. To prevent this siphon action, a vent must be coupled to the closet, somewhere between the seat of the plunger and the point where the closet is connected with the soil-pipe.

If the closet is conected with the soil-pipe without the intervention of a siphon-trap, the breaking of the seal in the overflow would put the occupants of the house in direct communication with the soil-pipe, as the covers to the plunger-chamber are not intended to have their joints air-tight, except in one or two instances. In selecting a closet of this class, care should be taken to get one with a plunger-chamber at least enamelled, while a bowl and plunger-chamber in one piece of earthenware is the best. The mechanism should be simple, the plunger-chamber small, and with a top or cover that can be easily removed.

The closet should be supplied with water from a cistern or tank, and should be admitted to the bowl through a flushing-rim. There should be a branch-supply for washing the plunger-compartment and plunger, admitted through a ring or sprinkler.

A hollow plunger simplifies the question of the overflow. It is necessary for the plunger to have a vulcanized-rubber ring or flange around its bottom.

The closet must have a siphon-trap under it, with a coupling or hub at its crown, to which a pipe may be soldered, or into which a pipe may be caulked; the vent must not be less than two inches in diameter.

SECTION V.

HOPPER-CLOSETS.

UNDER this heading I have thought it best to group all closets that have a simple bowl and no mechanical device other than a water-seal between them and the soil-pipe. This branch of the subject may with advantage be treated under the two types of long and short hopper-closets.

The long hopper is connected with a trap which is placed beneath the floor, while the short hoppers have their trap, which is often formed in one piece with the bowl, placed above the floor.

Long Hopper-Closets.—This type of closet being apparently the simplest form, we would naturally expect to find it the earliest kind put into use, but although this may have been the case there is no evidence to that effect; and if it was so, it was considered until recent years too simple to put on record.

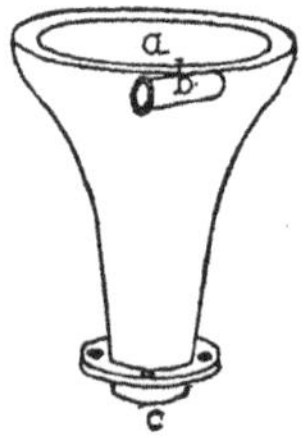

Fig. 160.
Philadelphia Hopper.

The opportunity for invention in this class is limited to the shape of the closet, the shape and position of its water-supply device and flushing-rim, or to the manner adopted for ventilating it.

Long hopper-closets are manufactured by all the firms making plumbers' supplies, and may be obtained of iron, plain, painted or enamelled, as well as in solid earthenware or porcelain.

The most common form is what is called the Philadelphia hopper. This has a bowl at the top, which suddenly changes to a diameter about the size of the soil-pipe. This closet is the cheapest and most objectionable form of its class. The sides are liable to become foul, the water being introduced in such a way that it can then flow around the bowl in a stream insufficient either to cleanse the sides of the bowl, or to empty the trap properly. The trap is connected with the bowl just below the floor line.

Foley's Closet.—A good type of an improved form of long hopper-closet was invented in 1883, by J. Foley. This closet as well as other better forms of the class are straight or nearly so at the back, so fæcal matter will drop directly into the mouth of the trap. The water-supply is conducted through an annular branch, formed in

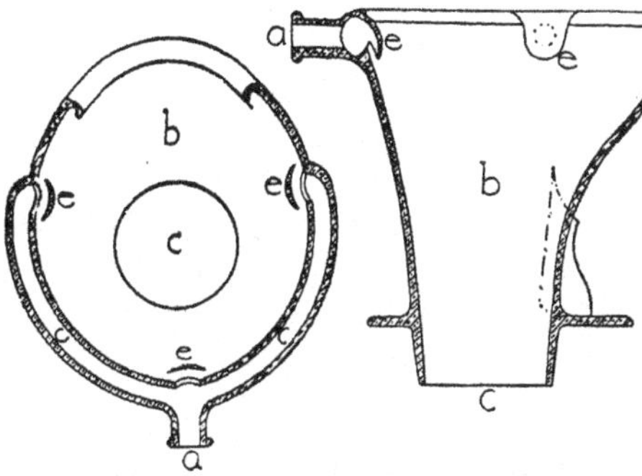

Fig. 161. Fig. 162.
Foley's Closet.
a, Inlet for water. *b*, Bowl. *c*, Outlet. *e*, Fans.

one piece of earthenware with the bowl. This conductor extends about half-way around the top of the bowl. The water is admitted to the bowl through three openings, one being at the back and one on each side. The water is distributed over the surface of the bowl by means of fans, instead of a flushing-rim, as is the case in other closets of its form. I think a good flushing-rim washes the bowl more thoroughly than any arrangement of fans.

Demarest Long Hopper.—The long hopper-closet furnished under this name by J. L. Mott & Co., is either round or oval in shape, and has one of Demarest's patent flushing-rims, the same that is used on the Hygeia closet (Fig. 139). This is in one piece of earthenware, and forms a very neat and effective closet when supplied, as it is intend to be, by a tank or cistern.

Myers's Niagara Closet.—A. G. Myers, of New York, manufactures a closet of this type and class. This hopper has Myers's patent zigzag flushing-rim, described in connection with the same firm's "All China Closet" (Fig. 156). It is useless to enumerate the many different closets which are similar to the ones named. All the prominent firms in England and America manufacture closets which vary from the ones illustrated only in the mode of flushing, or in a slight variation in their form. The long hopper closet is the best form for exposed situations where it is necessary to place the trap some distance below ground, according to the severity of the climate. It is necessary in a closet of this kind to have the supply-cock placed at the same distance below the surface, opening it by means of a long key or rod.

Fig. 163.
The Niagara Closet.
a, Bowl. *b*, Outlet.
c, Flushing-rim.

Maddock's "Wash-Out" Closet.—The Maddock's "Wash-Out" closet is of the long hopper class, and has a projecting lip formed on the upper and front part of the rim, to better adapt it for use as a urinal.

Fig. 164.
Maddock's Hopper.
a, Bowl.
b, Flushing-rim.
d, Outlet.
e, Projecting lip.

Rhoads's Closet.—The Rhoads long hopper closet was invented in this country in the year 1877. This closet is intended to be used without either a wooden casing or seat, in this manner avoiding the accumulation of, or collections of urine on, in, or behind the wood-work. A closet arranged in this manner is excellent for use in hospitals or prisons. In the first case the wood-work might absorb and communicate infectious diseases, in the latter case the wood-work would become foul from careless usage.

A large number of long hopper closets have vent-pipes attached to the lower part of the hopper just before it is connected with the soil-pipe or trap. Among closets with the vent, may be mentioned the "Worcester" closet, and I have been informed by

Mr. O. W. Norcross, that it was in use in the city from which it takes its name a number of years previous to the year 1876. In these closets the arrangement of the vent is for the same purpose, and in principle they seem to be identical. The receiver is a decided objection in the Smith closet, acting as a collector of filth: it is almost as objectionable as a receiver under a pan-closet.

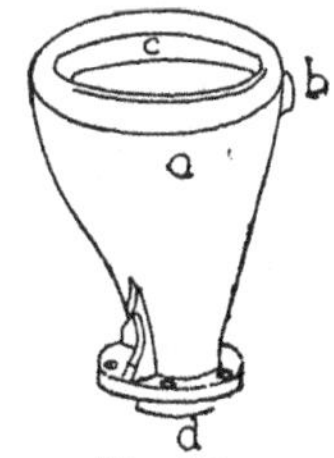

Fig. 165.
Rhoads's Hopper.
a, Bowl.
b, Supply.
c, Seat and flushing-rim.
d, Outlet.

The intention was to use these closets without a trap, in which case the vent near the soil-pipe, as in the Worcester closet, would be advantageous; but they are also intended to act as a vent for the room: for this purpose the Smith closet would have a slight advantage.

The Phillips Closet.—In 1882, a closet was invented which is a combination of the Worcester and Smith closets. The bowl is formed separately to set into a small hopper. The vent-pipe starts near the floor, runs along the side of the hopper, and has an opening into it near the bottom of the bowl. This

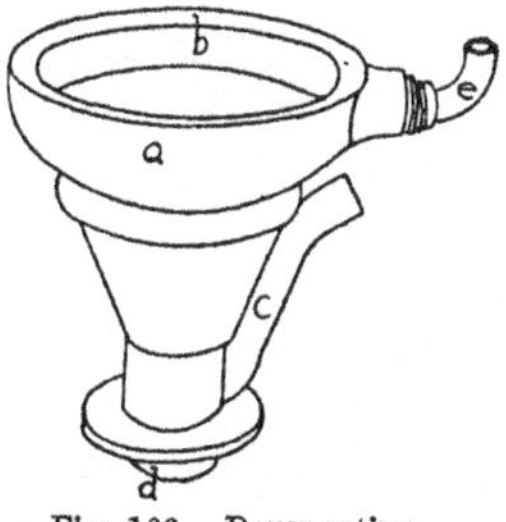

Fig. 166.—Perspective.

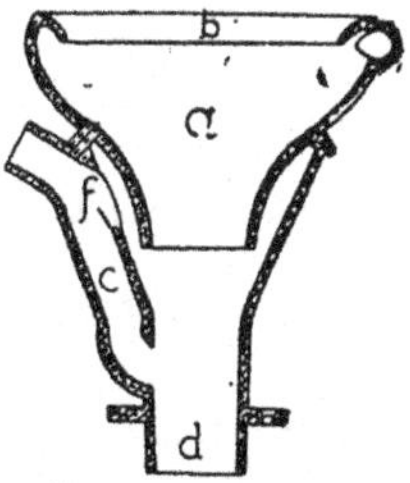

Fig. 167. Section.

Phillips Closet.

a, Bowl. *b*, Flushing-rim. *c*, Vent-pipe. *d*, Outlet. *e*, Supply-coupling. *f*, Upper opening into vent-pipe.

closet instead of having a metal hopper into which the bowl is set with the same objectionable putty-joint described in connection with the pan-closets, has the bowl, hopper and vent-pipe made in one piece of earthenware.

Watson's Closet.—Another closet of this kind, designed by Watson, is so arranged, by curving the outlet from the bowl, that the trap and its contents will be out of sight. This closet has a vent let into socket for a vent-pipe just before it joins the trap or soil-pipe. The bowls of this closet are made oval or square, and are much larger than this portion of the closet usually is. They can thus be used as slop-hoppers or urinals to better advantage than if the bowl were small.

There is no objection to using a vent-pipe in the position shown in the preceding closets; and it might be found advantageous in

Fig. 168.—Oval Bowl.

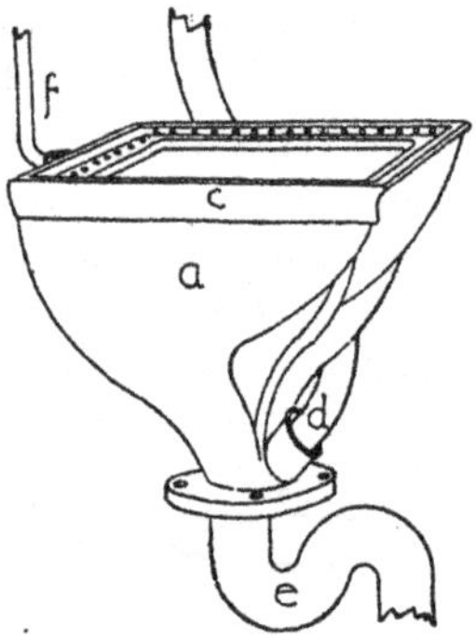

Fig. 169.—Square Bowl.

Watson's Closet.

a, Bowl. *b*, Vent. *c*, Flushing-rim. *e*, Trap. *f*, Supply.

taking off odors from the room, or odors created while some one was at stool. It must be remembered that a vent-pipe can never be a substitute for a proper siphon trap under the closet. Patentees sometimes claim that there is no further necessity for a trap, as the vent-pipe opening into a hot flue would carry off all noxious gases generated in what Mr. Eardley F. Bailey-Denton properly calls the inside sewerage system of the house.

The purpose which a vent-pipe of this kind is intended to fulfil—to carry off odors from the closet bowl or room—would be better served by an annular ring with slots in it, extending around the top of the bowl, and connected with some heated flue or pipe, and made either of metal separate from the closet, or of earthenware formed in one piece with the bowl.

Fowler's Closet.—Baldwin Latham describes a novel form of long hopper closet in the one designed by Fowler. The novelty in this closet consists in the trap being connected with the drain-pipe from the yard, sink, wash-basin, or other fixture.

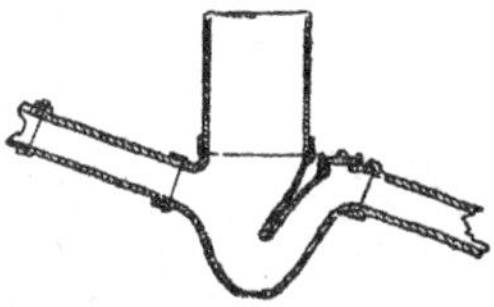

Fig. 170.—Fowler's Closet. *a*, Hopper. *b*, Trap. *c*, Drain. *d*, Inspection-cover.

In this manner only waste-water could be used in washing matter from the trap. Under some circumstances, in villages or country places a closet of this kind might be found useful. I consider the better long hopper-closets preferable to any other class or form, except the short hopper, and there are only two reasons why the short hopper is to be preferred. It has less distance between the bowl and trap, and thus less surface to become foul.

It appears to have been conclusively proved by the experiments of Messrs Edw. S. Philbrick and E. W. Bowditch made under the auspices of the National Board of Health, that the momentum of the water created by the distance between the top of the bowl, and the water in the trap was sufficient to force sufficient water out of the trap to break the seal. The vent-pipe (2") seems to have very little effect in preventing loss of seal by momentum, which must not be mistaken for siphon action. So it can be seen that a closet of this kind might lose its water-seal, and the party using it be none the wiser. The trap is such a distance from the closet seat, that it is impossible to see whether the trap has lost its seal or not.

SHORT-HOPPER CLOSETS.

This type may be conveniently treated by grouping together those which have a bottom outlet in distinction from those which have a side outlet.

Mr. W. P. Gerhard has seen proper in his book,* to classify a number of the short-hopper closets, some of which have their outlet

* *House Drainage and Sanitary Plumbing.* Wm. Paul Gerhard, New York, 1882.

at the bottom and others at the side, under the name of "Wash-out Closets." I think it simplifies the matter to place them where they certainly belong, in the class of hopper-closets. Mr. Gerhard himself says: "A wash-out closet is in fact only a modified and improved form of hopper."

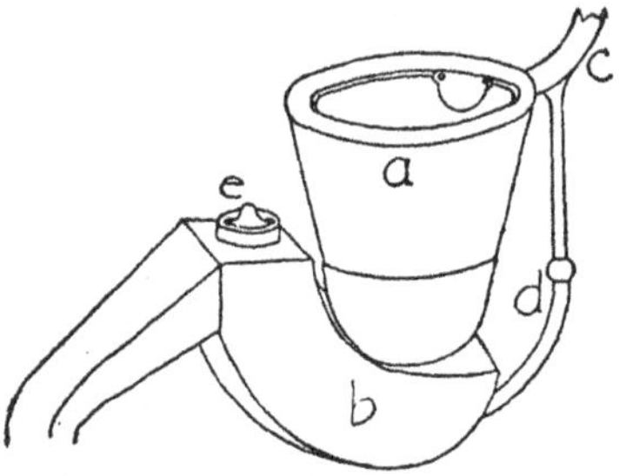

Fig. 171.—Thomas Smith's Closet.
a, Bowl. *b*, Trap. *c*, Supply-pipe. *d*, Branch to bottom of trap from supply-pipe. *e*, Inspection-hole.

Thomas Smith's Closet.—The earliest patent for a closet of this type was issued by Great Britain to Thomas Smith, in 1842. The patentee claims as a novelty the branch from the supply-pipe (the closet being supplied from a tank), which is connected with the bottom of the trap. This branch from the supply-pipe is intended to thoroughly wash out any matter which might be in, or would otherwise remain in the trap. In this closet the trap, instead of being circular in cross-section, is rectangular. It has an inspection-cover on the crown of the trap, and the water is spread over the bowl by means of a fan.

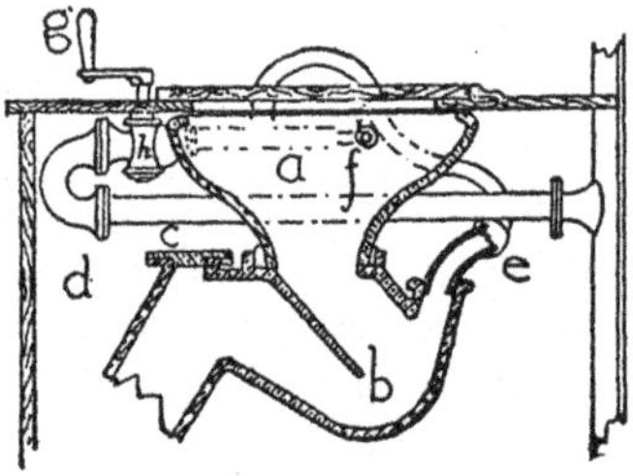

Fig. 172.
French Short-Hopper Closet.
a, Bowl. *b*, Trap. *c*, Inspection-hole. *d*, Supply-pipe. *e*, Branch from supply-pipe connected with bowl. *f*, Opening into bowl. *g*, Lever opening supply-valve.

French Short Hopper.—F. Liger* tells us that this closet (Fig. 172) was designed in Paris, in 1872, for use in public places, being modelled according to English patterns. He states "that the gases from the sewer are cut off by a siphon-trap." In public places where this closet had been introduced, it was made to operate by opening the doors. The closet is arranged so it can

* *Fosses d'Aisance, Latrines, Urinoirs, et Vidanges.* F. Liger, Paris, 1875.

be supplied with water, either by means of a supply-valve or from a tank. The peculiarities of this closet consist in the branch from the main supply-pipe for flushing the trap, and in the partition of the trap being so placed as to form a screen, and hide, rather than get rid of, any disagreeable matter that might possibly remain.

Hellyer's "Vortex" Closet.—S. S. Hellyer, of London, received patents from Great Britain in 1878, and from the United States in 1880, for a short hopper closet which he has named the "Vortex." The novelty of this closet is based on the shape of the bowl and flushing-rim. The flushing-rim is so arranged as to throw a strong

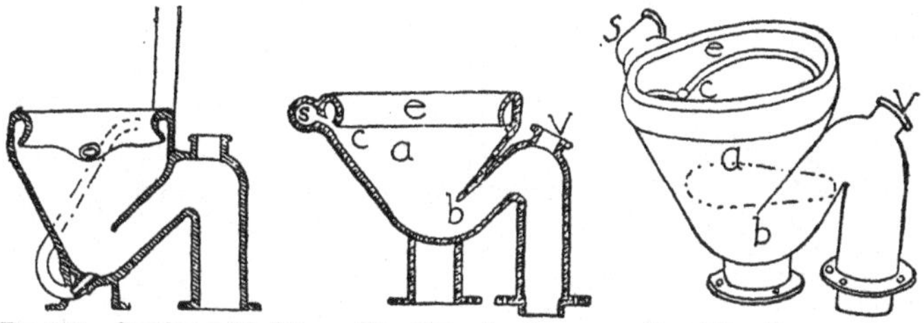

Fig. 173.—Section with Jet. Fig. 174.—Section. Fig. 175.—Perspective.
Hellyer's "Vortex" Closet.

a, Bowl. *b*, Trap. *c*, Opening for jet of water. *e*, Flushing-rim. *s*, Supply. *v*, Vent.

jet of water directly into the bottom of the bowl, this being also the bottom of the trap, by this means forcing whatever water or foreign matter there is through the trap into the soil-pipe. It will be seen by the illustration (Fig. 173) that it is intended in some cases to have a branch from the main supply-pipe feeding the bottom of the trap, as described in the Smith closet (Fig. 171). There are openings in the flushing-rim around the sides of the bowl, by which the surface of the bowl is washed. The trap is properly vented. This branch connecting with the bottom of the trap is unnecessary in the better forms of short-hopper closets, where they are flushed from a tank, through a properly-arranged flushing-rim.

Wm. Smith's Closet.—The "Siphon-Jet" closet, manufactured in San Francisco, and patented in 1876 and 1878 by Wm. Smith, also depends upon the branch from the supply-pipe to cleanse the trap and force the waste matter into the soil-pipe. This closet is furnished in one piece of earthenware, or with an earthenware bowl and iron trap. It is connected with a tank, or may be connected directly with the supply-pipe, the flush being governed by a patent valve.

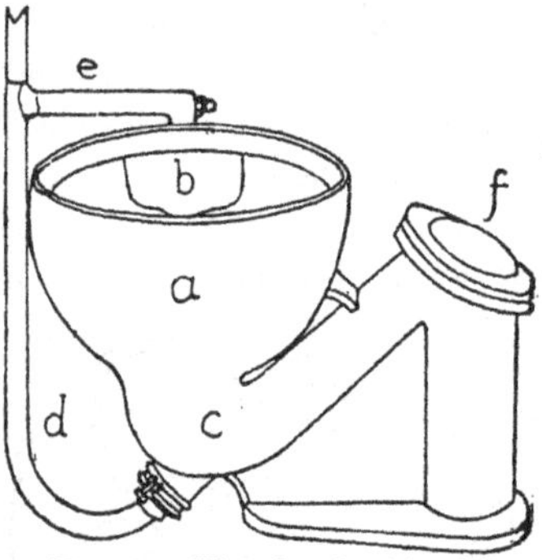

Fig. 176.—Wm. Smith's Closet.
a, Bowl. *b*. Fan. *c* Trap. *d* Branch supply-pipe. *e*, Supply-pipe. *f*, Inspection-hole.

Mr. J. Pickering Putnam, in 1884—within the last month—has patented a closet with a jet supply to the trap.

The jet is large. It forms a part of the tank, and the water is held in position by a vacuum formed in the top of the tank itself. It claims to be noiseless, as the water enters below the water line. I should think this large receptacle would become foul by the water from the closet. From the illustrations it appears complicated and to have many corners and crannies to collect dirt.

English Closet.—Baldwin Latham describes a short-hopper closet in which the flushing-rim is connected with the bottom of the trap. He says: "this form of closet with bye flushing-pipe was introduced to remedy the defect, which, as already referred to, attends the hopper-closet, that is the cleansing of the trap after the use of the closet." This evil mentioned by Latham has been remedied by properly arranged and devised water-supply and flushing-rims, which are now manufactured as a part of the better short-hopper closets.

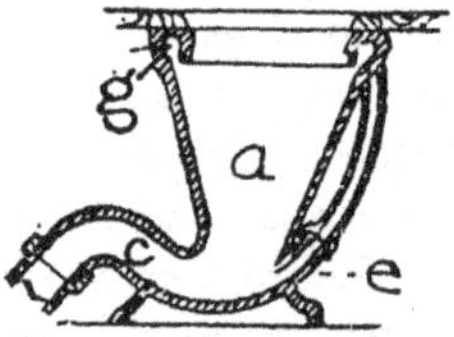

Fig. 177.—English Closet.
a, Bowl. *c*, Trap. *e* Branch-supply. *g*, Flushing-rim.

Closets manufactured as described, with a branch from the supply-pipe connecting with the bottom of the trap, would be peculiarly susceptible to the evil which it is claimed sometimes takes place when the supply-pipe is attached directly to the bowl of the closet, a valve only intervening; viz., the liability of foul air or filthy water being drawn from the closet when a faucet is opened on some floor below, the supply-valve of the closet being open at the same time. For the reason stated care should be taken always to connect closets of this kind with a separate tank or cistern when—and they are simple and useful—they are used.

Hellyer's "Short Artisan Hopper."—This closet is composed of two pieces of earthenware, the bowl and trap, which are joined together by means of clamp-screws, putty or cement being put into the joint. The bowl of this closet is oval in shape, with its back edge nearly straight, so the bowl need not become foul by fæcal matter dropping on it. This closet has a flushing-rim which serves its purpose well, and the trap is properly vented at the crown. In

Fig. 178.

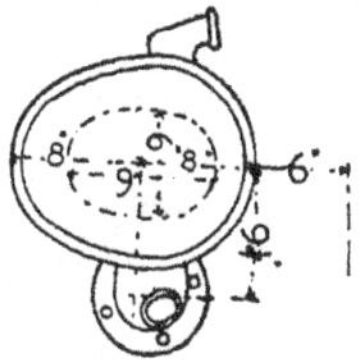

Fig. 179.

Hellyer's Short Artisan Hopper.

connection with his patent flushing-rim, Hellyer arranges an annular rubber band, in a groove made for the purpose on the top of the flushing-rim. The seat of the closet is intended to fit down on this band and thus prevent smells or cold air from coming into the room from beneath the seat. It is the custom in England to have the "tell-tale" pipe from the closet-safe run through the wall into the outer air. A draught through this pipe might become disagreeable to a person occupying the seat of the closet. This closet has an opening at the crown of the trap, to which a vent must be attached, and in use it is simple and effective.

Fig. 180.—Flushing-rim. *i*. India-rubber ring. *c*, Flushing-rim.

Demarest's Short-Hopper Closet.—The J. L. Mott Iron Works manufacture a short-hopper closet which in appearance and in its manner of working is almost identical with the Hellyer short hopper. The novelty consists in the manner of connecting the bowl with the trap, and the trap with the vent-pipe, and in the Demarest patent flushing-rim (see Fig. 180). The top of the trap and the bottom of the bowl are both encircled by a projecting ring or flange, over and around which a metal collar fits. This collar when bolted makes a very strong joint. The space between the collar and the joints has to be filled with putty or cement. The coupling for the vent-pipe is a two-inch zinc-coated iron quarter-bend, which has a hub on the free end, into which the spigot of the vent-pipe can be caulked. If a lead vent-pipe is to be used, a brass coupling must be connected with the trap instead of the iron one described.

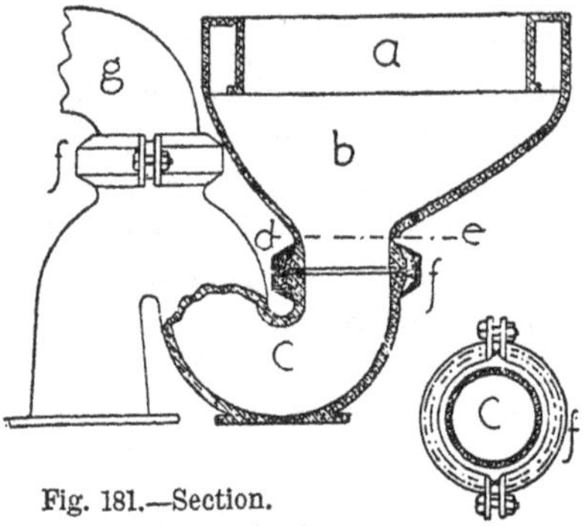

Fig. 181.—Section.
Demarest's Short-Hopper Closet. Fig. 182.
Section through line *d-e*.
a Flushing-rim. *b*. Bowl. *c*, Trap. *g*, Vent-pipe. *f*, Collar or clamp.

Roe's Closet.—Baldwin Latham, in his work on sewerage, illustrates two short-hopper closets. Mr. Latham states that in this closet exists the first instance of a vent-pipe being properly placed on the trap. The reader of the foregoing pages will probably

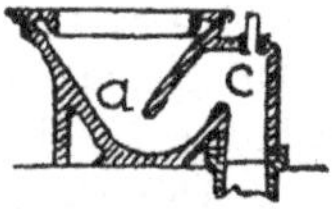

Fig. 183.
Roe's Closet.
a, Bowl. *b*, Trap.

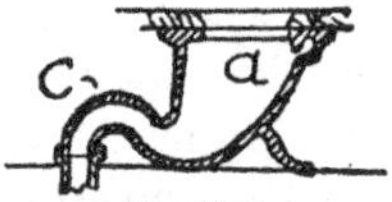

Fig. 184.
Straight-Back Hopper.
c, Vent.

recollect that the vent to the trap was used or invented in connection with the Wilkins closet forty years ago. Figure 184 is shown as an example of this class, in which the back is made straight, so fæcal matter will drop directly into the trap, and not adhere to the sides of the bowl.

Carmichael's "Wash-down" Closet. Mr. W. P. Buchan, in 1879, invented a closet of this class, which he informs us is furnished in one or two pieces of earthenware, the joint being made above the water-line.* This closet is supplied with a flushing-rim to wash the sides of the bowl, and also with a jet of water which strikes the water in the bowl so as to force it through the trap. The trap has an inspection-hole on top. This closet seems to have deservedly met with success in England, where there seems to be a tendency, which should be followed in this country, to adopt short-hopper closets in preference to all styles or patterns. Mr. Buchan calls the short-hopper closets which have an outlet at the bottom wash-down closets, while those which have an outlet from the side he calls wash-out closets.

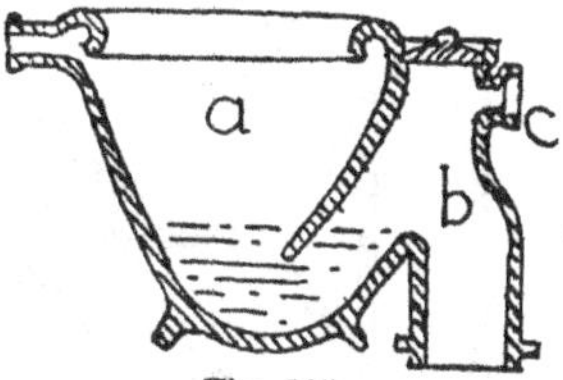

Fig. 185.
Buchan's "Carmichael" Closet.
a, Bowl. *b*, Trap. *c*, Vent.

A. G. Myers's "Niagara" Closet No. 2.—A short-hopper closet has been quite recently introduced into the market by A. G. Myers, of New York. This closet is made in one piece of earthenware, or it is furnished with a part of the trap separate from the bowl. It is flushed by carrying the water-supply through or against a fan. The bowl of this closet is waved or corrugated in circles parallel to the floor. The idea is that the water, coming in contact with the inequalities of the surface, will wash the bowl more thoroughly than would be the case if it were smooth. It appears to me that the under side of the wavy surface would not be washed at all, and if this is the case, more harm than good would be done by making a bowl in this manner.† The trap in this closet has a hub for a vent-pipe, which is placed, in accordance with the suggestion of Hellyer, on the side of the trap nearer the soil-pipe than is usual. This position is intended to prevent foul matter from being forced up into, and lodg-

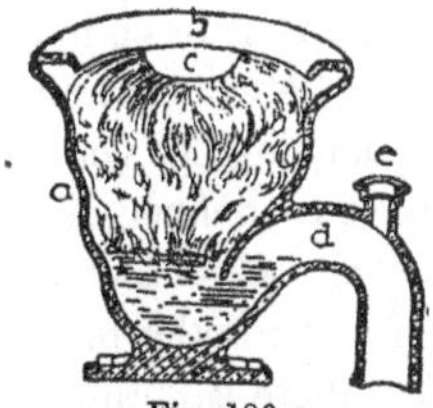

Fig. 186.
"Niagara" No. 2.
a, Corrugated bowl.
b Fender. *c*, Pan.
d, Trap. *e*, Vent.

* ***Plumbing.*** W. P. Buchan, London, 1883.

† Since the above was written, Mr. Myers has changed the form of this closet, and put it on the market in the form of Hellyer's short hopper (see Figs. 178–179).

ing in the vent-pipe. It is claimed that the vent-pipe is sometimes made useless in this manner.

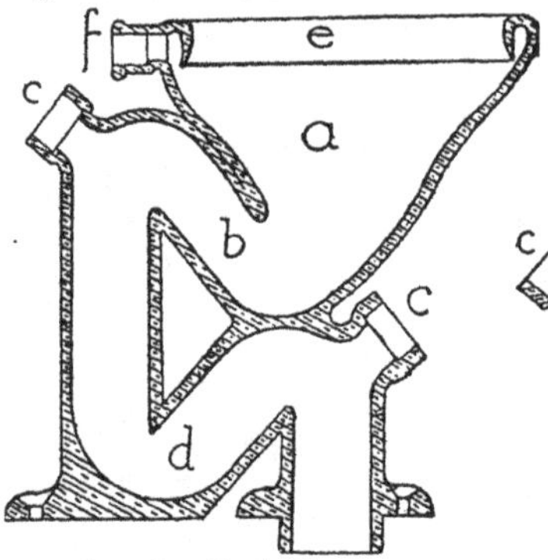

Fig. 187.—Single Trap.
a Bowl. *b*, Trap No. 1.
d Trap No. 2. *c*, Vents.
e, Flushing-rim. *f*, Water-supply inlet.

Owens's Closet.—Henry Owens, in 1878, invented a closet which is made in one piece of earthenware, and is designed so as to have either two traps or a trap and an offset between it and the soil-pipe. The traps, and also the offset have openings for vent-pipes. The closet with the double trap would have a confined air-space which could only be ventilated by an inlet as well as an outlet pipe for the air. I think as there is nothing gained by the double trap, the closet with the trap and offset is to be preferred.

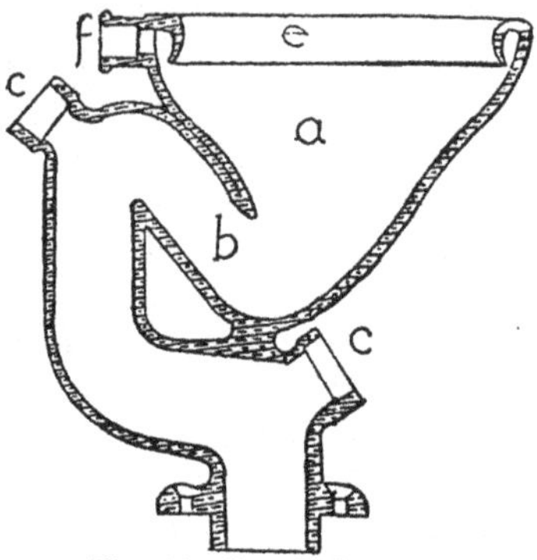

Fig. 188.—Double Trap.
Owens's Closet.

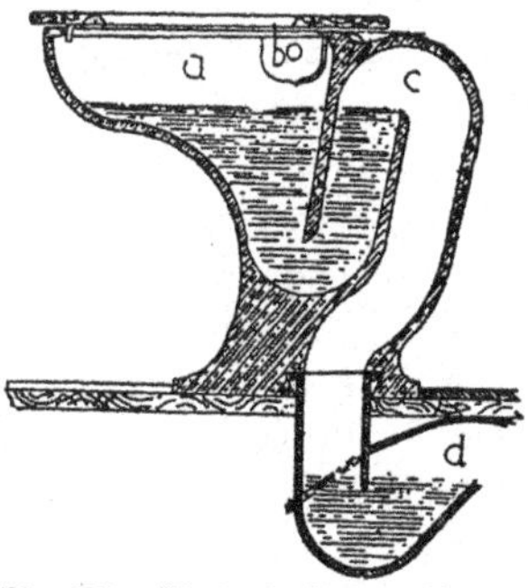

Fig. 189.—Waring's Siphon Closet.
a, Bowl. *b* Pan. *c*, Siphon.
d, Trap No. 2.

Waring's Siphon Closet.—This closet, which was designed by Mr. Geo. E. Waring, Jr., of Newport, R. I., has a very deep water-seal trap at the the bottom of the bowl. The closet is set over a weir, which is placed beneath the floor. The contents of the bowl are siphoned out every time the bowl is used. From its shape it is necessary that this closet should be filled to within a short distance of the top of the bowl, before the necessary siphon action which empties the bowl can take place. This excessive amount, or, more properly speaking, the height to which the water rises in the bowl, is exceedingly disagreeable sometimes. The weir below the floor is in shape or form very much like the objectionable D-trap. The first trap by reason of its siphon action cannot have a vent-pipe in the usual position. The closet is gracefully formed in one piece of white earthenware.

Boyle's "Tidal Wave" Closet.—In the year 1882 a closet with the double trap was invented by J. E. Boyle. An air-pipe runs from the space between the two traps, up and through the tank. This air-pipe is governed by a valve which opens only when acted on by a pressure from below, so this vent-pipe would not interfere with forming of the partial vacuum, which would be necessary before the closet could be emptied by siphon action. When two-thirds of the service-box has run off, the vacuum is broken and the remaining third fills the bowl. The bowl and its double trap is manufactured in one piece of earthenware. The lower trap is properly ventilated. The closet must be supplied from one of Boyle's patent cisterns or tanks.

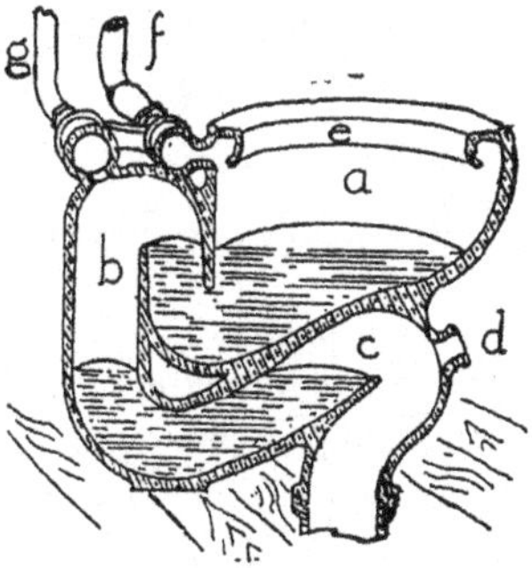

Fig. 190.—Boyle's Closet.

a, Bowl. *b*, Siphon. *c*, Trap. *d*, Vent. *e*, Flushing rim. *f*, Supply pipe. *g*, Air pipe.

William Bunting, of New York, received patents from the United States, in 1878, for a closet in which the siphon action is created by a piston which works in a chamber situated above the trap.

A. G. Jennings's Side-Outlet Hopper.—The first short hopper with a side outlet was invented in 1852, by A. G. Jennings, of London, in which he claims as novelties "the basin (bowl) and trap combined in one piece of earthenware, so formed that the basin is made to contain a quantity of water.....The escape may be from the face, side, or front. The basin may be of any shape to hold water in the bottom." In 1876 an improvement was made on this closet, for which a patent was issued. He says: "It has been found to have the disadvantage that the water tends to run around and around, instead of passing out in a stream." To remedy this defect, a rib, which is shown in the illustra- by a dotted line, was introduced. This rib causes the water to be thrown in a mass toward the outlet, carrying whatever waste matter there was in the bowl with it through the trap. An opening for a vent-pipe, which is not shown in the illustration, is now manufactured in the usual position on the crown of the trap. The water is supplied through a fan instead of a flushing-rim, and it is intended to be connected with a tank.

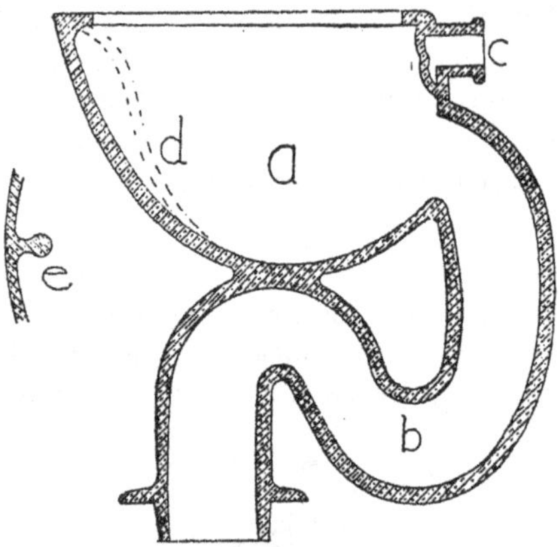

Fig. 191.—Jennings's Side-Outlet Hopper.
a, Bowl. *b*, Trap. *c*, Supply. *d*, Dotted lines showing where rib *e* was added in 1876.

Buckland & Rees's Closet.—The second closet of this class was invented in England in 1862, by Buckland & Rees. This closet has its fan so shaped that a portion or half the water entering the bowl is driven to each side, "instead of being driven with a spiral motion, as is usually the case, the greater part of the water being allowed to flow directly to the bottom of the basin, and thence through the aperture or discharge-pipe on the other side." This closet has its trap and bowl in separate pieces. The "National" side-outlet closet manufac-

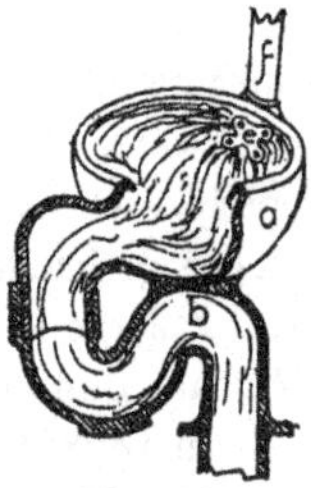

Fig. 192.
Buckland & Rees's Hopper-Closet.
a, Bowl. *b*, Trap.
c, Fan. *f*, Supply.

tured in this country is apparently a combination of the two closets which have just been described.

Hellyer's Side-Outlet Hopper.—I illustrate two forms of short hopper of the type under discussion, which were invented by S. S. Hellyer, of London. He acknowledges that he was unable to make closets of this kind act satisfactorily. The fæcal matter would strike against the sides of the outlet, and leave it in a foul condition. To obviate the tendency which the waste had to remain in the bottom of the bowl, the supply-pipe was divided into

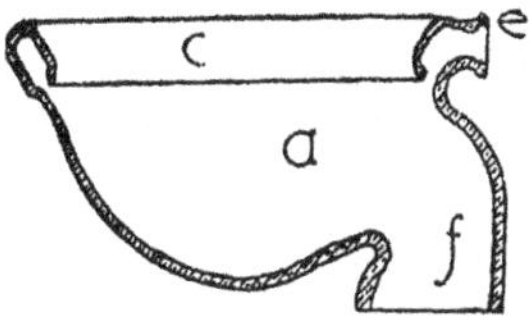

Fig. 193.

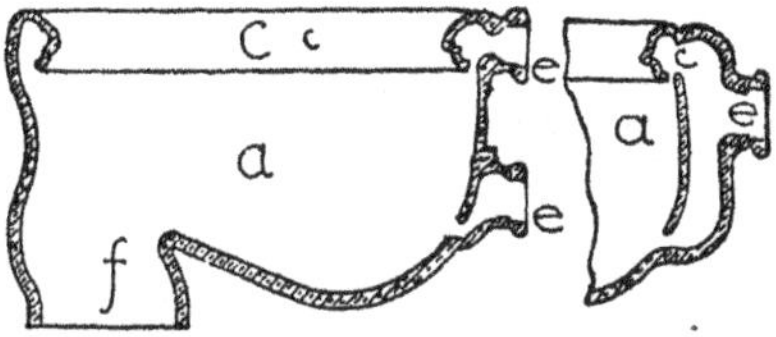

Fig. 194.

Hellyer's Side-Outlet Short-Hopper.

a, Bowl. *c*, Flushing-rim. *e*, Inlet for supply. *f*, Outlet.

two parts, one of which entered the bowl near the bottom, and in this manner forced the water and foreign matter through the outlet. Hellyer concluded that this particular form of hopper was not good for the purpose which it was intended to serve. The bowl is formed so it can be placed in an ordinary trap that is placed above the floor.

Bostel's "Brighton" Closet.—A closet of this class, which has recently been extensively used in this country and in Europe, was invented in 1877, by D. T. Bostel, of Brighton, England. This closet is made in one piece of earthenware, the bowl and trap being formed in a manner similar to the Jennings short hopper (Fig. 191), the difference being in the flushing-rim, inspection-hole, and in the vent-pipe. The supply-pipe is divided so as to enter the flushing-rim at two points. Flowing around the flushing-rim, the water enters the bowl through small holes, which are purposely made larger on the side of the bowl, opposite to, or furthest from the outlet. By this arrangement the greatest volume of water is thrown where it is most needed, and the matter in the bowl is carried into the trap, if not through it. There is an opening with a cover, to be used for the purpose of inspecting the mouth of the trap, and form-

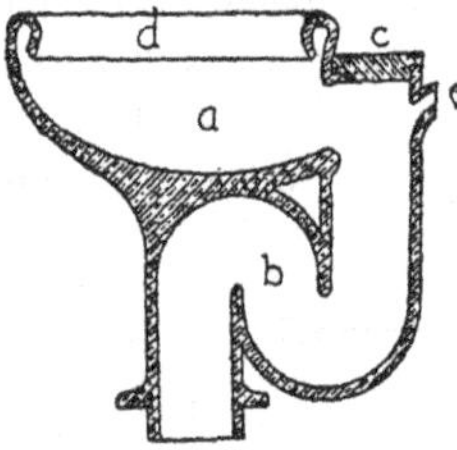

Fig. 195.

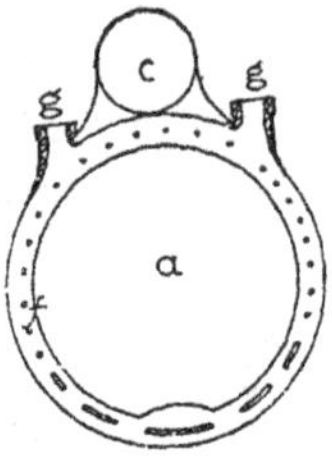

Fig. 196.

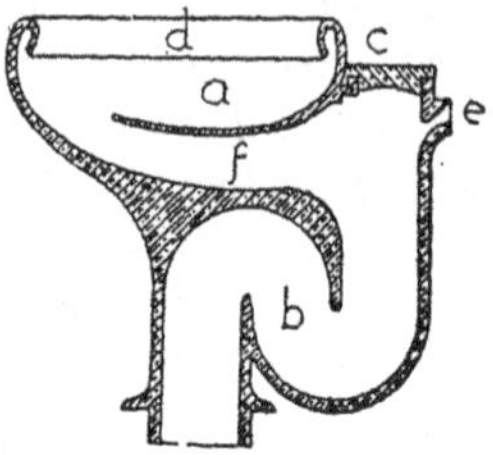

Fig. 197.

The "Brighton" Closet.

a, Bowl. *b*, Trap. *c*, Inspection-cover. *d*, Flushing-rim. *e*, Vent. *f*, Slots in flushing-rim. *g*, Double-supply.

ing a top to the side outlet. The opening for a vent-pipe is just below the inspection-cover. A vent-pipe in this position will only serve to carry off locally-generated gases, and must never be connected with the soil-pipe, as it would form a communicating duct between the inside sewerage system and the house. In this closet a vent-pipe to prevent siphonage, back pressure, and to carry off gases generated in the soil and other waste pipes, would have to be connected with the branch from the soil-pipe below the floor, as there is no opening for the purpose near the trap. There is sometimes a false bottom or screen manufactured in this closet. I can see no good purpose served by this screen, while it increases the space and surface between the bowl and trap, which is liable to be fouled by matter in its passage to the trap. This closet is arranged to receive its water from a tank.

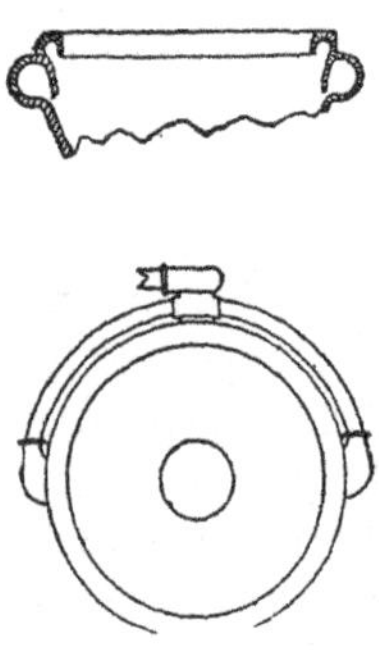

Fig. 198.
Stidder's
Branch Supply.

Stidder's Closet.—In 1878, one Stidder invented, and received patents from Great Britain, for a supply-pipe divided so the water could enter the flushing-rim at three points. This flushing device was intended for use with a short-hopper closet.

The Lambeth "Flush-out" Closet.—A closet of this form and class, which was designed and is manufactured by Doulton & Co., of England, is made in two pieces of earthenware. This closet has an inspection-hole so that the compartment over the top of the trap can be examined without difficulty. There is also an opening in the crown of the trap, which may either be used as an inspection-hole or as a place into which a vent-pipe can be inserted. The flushing-rim is similar to the one described in connection with the Lambeth "Trapless" closet (Fig. 61). The openings in the flushing-rim, opposite the side-outlet, are so arranged that a large body of water is thrown directly into the bottom of the bowl, and in this manner the bowl and trap are cleaned by the scour of the water. The J. L. Mott Iron Works, of New York, have recently introduced a closet of this kind into the market, which has an inspection-opening over the side outlet, and a brass coupling in the usual position on the soil-pipe side of the trap, to which a vent-pipe must be soldered or caulked. These closets are furnished in one piece of earthenware, or with an earthenware bowl and an iron trap, and are called the "Inodora" and "Purita."

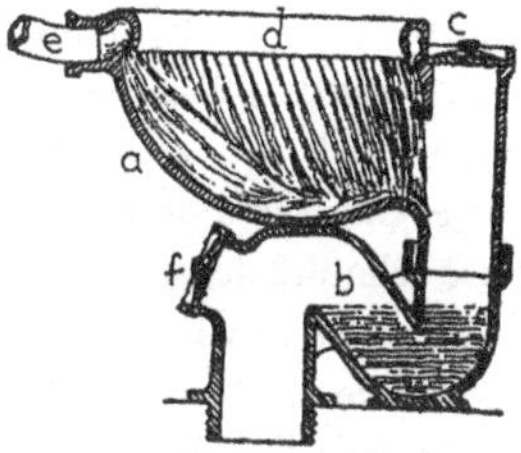

Fig. 199.
The Lambeth "Flush-out" Closet.
a, Bowl. *b*, Trap. *c*, Inspection-cover. *d*, Flushing-rim. *e*, Supply. *f*, Opening for vent.

DEDUCTIONS.

As stated before, *I consider the simple short-hopper closet, with a bottom outlet, the best form of water-closet in use at the present time*, when it is properly flushed from a tank, through a well-arranged flushing-rim, and has a trap properly ventilated. Although I have not found it necessary, a vent-pipe encircling the bowl and connected with a heated flue may sometimes be found advantageous.

In this form of closet we find one simple in its construction, effective in its action, immediately discharging the fæcal matter into the soil-pipe, and at the same time cleanly and healthful, having no concealed chambers or compartments to become foul and generate unwholesome gases. In fact, it leaves little to be desired as an improvement on water-closets. Closets with bottom outlet and of the short-hopper pattern, with effective flushing-rims, are manufactured by the majority of dealers in plumbers' supplies in the United States and Great Britain, and the French have also made an effort to manufacture and introduce them into Paris. The short-hopper closets of the form under consideration are recommended by such prominent authorities on sanitary plumbing as S. S. Hellyer,* of England; W. P. Buchan,† of Scotland, and Prof. P. Putyo‡ in a work recently published in Belgium. By a comparison with other forms which are and have been in use, the cleanliness, simplicity and effectiveness of the short hopper can easily be appreciated by those who are least expert on the subject. With the valve-closet we have the bowl and its usually faulty connection with the receiver. The valve itself and its seat, nicely adjusted with all necessary cranks, spindles, journals, springs, levers, weights, etc., according to the pattern of closet, are all liable to get out of order, and all parts which are within the receiver and the sides of the receiver are liable to become foul. Portions of the valve are formed of perishable rubber, the overflow is more or less complicated, and the trap, when one is used, is necessarily placed below the floor, with few exceptions.

The plunger-closet has its plunger, plunger-compartment, and, in all except a very few cases, floats aud supply-valves with overflows, all connected with the bowl, and between it and the trap.

In the classes just mentioned, all the described mechanism is used in place of the simple bowl and trap of earthenware. And with what gain? The speedy and effectual removal of waste matter is accomplished more quickly and thoroughly by the simple than by the complex closet. In the plunger and valve closet small particles of filthy matter and foul sediment collect in the compart-

* *The Plumber and Sanitary Houses.*

† *Plumbing.*

‡ *L'Hygiene dans la Construction des Habitations Privees,*

ments which are under the valve or around the plunger, and concealed from view. These particles decay and generate foul and in some instances poisonous gases.

The whole surface of the short hopper, except in the side-outlet form, is visible, and any particles that cling to the bowl are in plain view, and I think all will acknowledge that it is better to have such foulness where it can and will be cleaned off by a careful housekeeper, than to have it out of sight, where it will remain and undergo decay; fair without, but foul within. When these closets (the simple short hopper) have their seats in the proper position, the tank placed at a sufficient height, between seven and eight feet, one-and-a-quarter-inch supply-pipe, which must not be diminished by bends, I have never known them to fail to wash the bowl perfectly whenever flushed. The plunger and valve closets are supposed by many to form a mechanical seal between the bowl and the soil-pipe, but this is not the case with the closets of this class which are in the market, as the overflow only interposes a water-seal, except in the closets mentioned in preceding pages, which is the same seal as the trap of the short hopper. Although the experiments of Dr. Fergus, of Glasgow, and Doremus, of New York, have apparently proved that gases in a concentrated form will pass through a water-seal trap, nevertheless I think it can be said without exaggeration, that the majoity of sanitarians have accepted the experiments of Dr. Neil Carmichael, of Scotland, which have since been confirmed by Dr. Wernich, as conclusively proving that "water-traps are therefore, for the purpose for which they are employed, that is for the exclusion from the house of injurious substances contained in the soil-pipes, perfectly trustworthy;" in other words, organic germs will not pass through a water-seal trap. With the traps vented and the soil-pipe open, we need have no fear of gases passing through the trap, provided the water seal remains intact.

The advantage which a short hopper has over a long hopper has been mentioned in connection with the description of the latter type, as having less surface to become foul between the seat and the trap, and the possibility which exists of the trap under the long hopper losing its seal by the impetus attained by the water. The advantage which this form has over the hopper with the side outlet is having the mouth of the trap always visible, and being without the concealed space formed by the side-outlet between the bowl and trap, which should always be avoided. It is needless to say that the pan-closet, as it is usually manufactured, with its container which becomes almost filled with filth, the faulty connection between bowl and container, which allow gases generated to enter the room, its imperfect flush and leaky supply-valves, is not to be compared with the short-hopper closet.

SECTION VI.

MISCELLANEOUS CLOSETS.

THE two closets of which the following is a description, do not belong to either of the classes which have been enumerated; at the same time they appear worthy of description.

The first is a closet of German pattern, described by Baldwin Latham in his work on sewerage.

German Closet.—This closet consists of a side-outlet hopper, with a siphon trap above the floor, in the bowl of which there is a tilting-basin that can be filled with water. The inside basin conceals the outlet from the bowl into the trap. When this basin is lifted, or tilted, its contents are discharged through the trap into the soil-pipe.

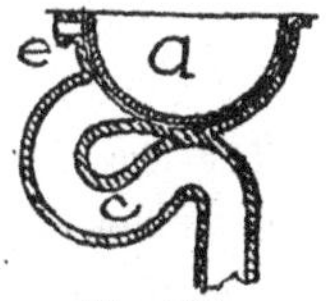

Fig. 200.
German Closet.
a, Tilting-basin.
c, Trap.
e, Supply.

This closet in its manner of operation is almost exactly like the tilting wash-basins patented and manufactured by J. G. Jennings, of London, and I should think the space between the two basins was liable to become very foul.

The second closet that belongs to the miscellaneous class, which I have selected for description, is one invented by John Tylor & Son, of London, in the year 1873.

This closet has a large receiver (in every way equal to, and as objectionable as the receiver under the ordinary pan-closet), into which the bowl projects.

The projecting outlet from the bowl turns at a right angle with itself after entering the receiver, and over its outlet is fitted a horn-like valve which revolves on a short arm and has its upper end open. This valve has a rubber band stretched over its larger end, where it fits against a metal seat that is clamped to the outlet of the bowl. In this manner the water-closet bowl is kept filled with water to the height of the small end of the horn. The overflow, which, according to Mr. E. F. Bailey-Denton, is now supplied with a ball-valve, takes place through the horn.

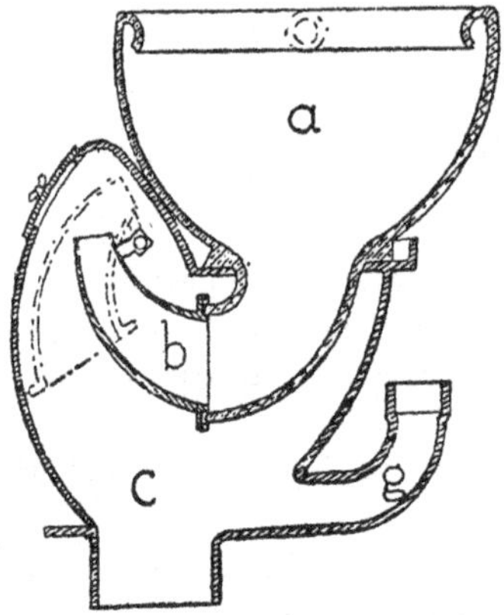

Fig. 201.—Tylor & Son's Closet.
a, Bowl. *b*, Horn-valve. *c* Receiver. *g*, Bath-waste.

There is an inlet cast on the receiver of this closet into which bath or water-basin wastes may be caulked.

Neither of the above described closets probably will ever come into extended use, as they are inferior to most plunger and valve closets, and for this reason far inferior to the class of hopper-closets.

LATRINES.

The class of which the following is a description is in reality a combination of closets, rather than a single one, and is always intended for use in public places, such as depots, schools, hospitals, barracks, asylums, or any place where the usual closet will most probably be neglected or abused, and where a janitor can take charge of flushing and emptying them; the plug, valve, or tank, which forms the mechanism of the closet are under lock and key, and accessible only to the janitor in charge of the building.

The Lambeth Improved Latrine.—Doulton & Co., of England, manufacture a latrine that consists of a series of stone-ware bowls connected with each other and the main drain by suitable stone-ware pipes. The plunger is a simple hollow one that is formed by an iron tube. The plunger forms the overflow and has a perfect seat over the siphon-trap. The plunger is lifted by a handle that is kept under look and key by the janitor.

The bowls are filled from one cock. When it is necessary to

empty the latrine, the handle is raised and the outlet pipe (5" diameter) is opened, when the contents of the whole row of bowls and connecting-pipes will be discharged through the trap into the soil-pipe, in a few moments.

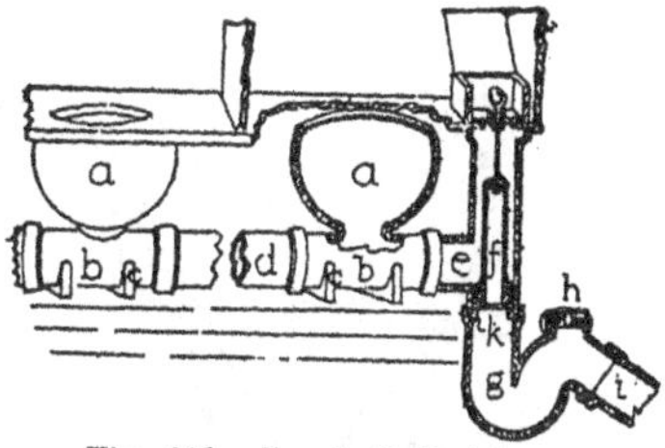

Fig. 202.—Lambeth Latrines.
a, Bowls. *b*, Connecting-pipes jointed to bowls. *d* Connecting-pipe. *e*, Plunger-chamber. *f*, Plunger. *g*, Trap. *h*, Inspection-cover.

The trap of this latrine is supplied with an inspection-hole, and it should also have, when used, an opening for a vent-pipe on the side of the trap next to the soil-pipe.

Jennings's Patent Latrines.—These latrines have been extensively used in public buildings in England. They are composed of bowls made on a special spherical pattern, connected with each other by means of short pipes, the flanges of which are bolted together. The bowls rest on short legs, so the whole system is raised above the floor. The bowl is divided in hemispherical pieces. This form and arrangement is convenient for shipping, as the upper half will set into the lower one. The bowls are furnished either in highly glazed vitrified stone-ware or enamelled iron. The plug is one of Jennings's hollow plungers opening over a simple siphon-trap. The plunger should be raised once a day, and oftener if ample water can be obtained. The water-cock as well as the plunger should be in charge of the janitor.

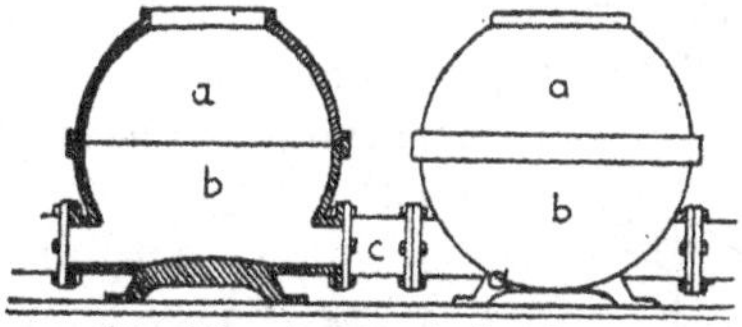

Fig. 203.—Jennings's Latrines.
a, Upper part of bowl. *b* Lower half. *c*, Connecting-pipe. *d*, Feet.

Tylor's Latrine.—In the year 1873, Tylor & Son, of London, received patents for a latrine of a peculiar form. The bowl is formed so as to set into and rest upon a receiver. This receiver has on each side at the bottom a projecting spigot that could be cemented or caulked into the bell of a pipe. In this way a continuous line of latrines can be formed. The bowl, which is oblong,

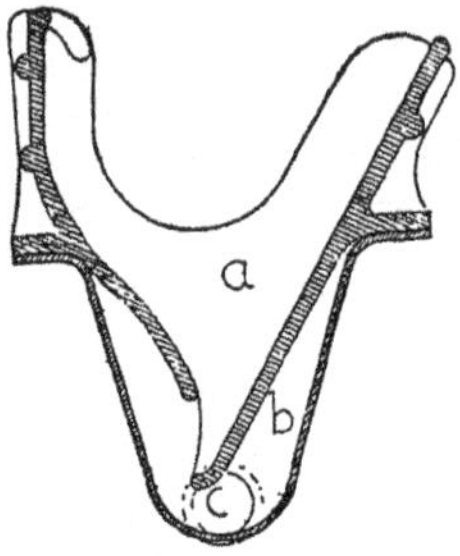

Fig. 204.

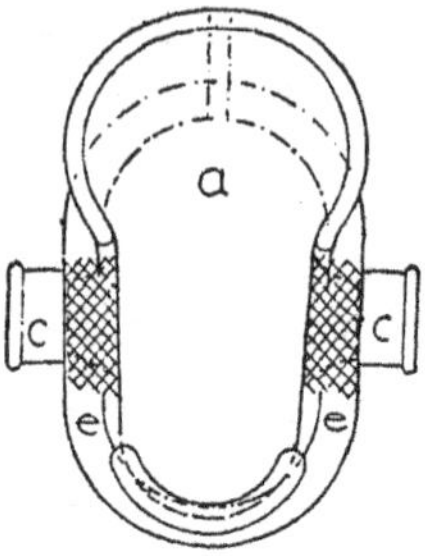

Fig. 205.

Tylor's Latrine.

a, Bowl. *b*, Receiver. *c*, Spigot ends.

has foot-rests on each side, and is intended to be used by parties crouching, no seat being provided. The patentee says they were invented for the foreign market. In public parks where water-closet seats would be liable to abuse, or where they might communicate disease, latrines of this or a pattern to accomplish the same end would be found useful.

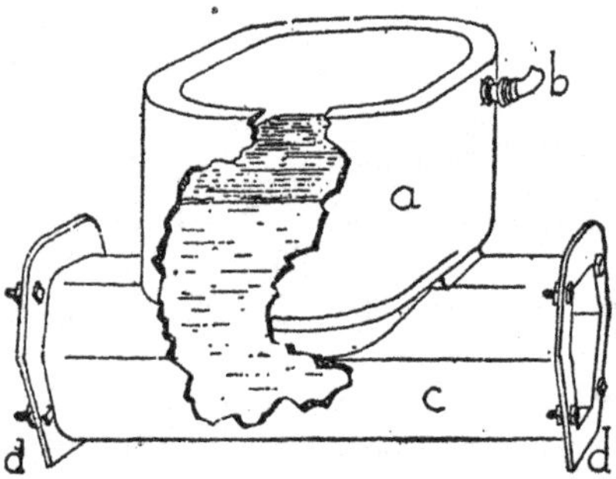

Fig. 206.—Mott's Latrine.

a, Bowl. *b*, Supply. *c*, Connecting-pipe.

Mott's Latrines.—The J. L. Mott Iron Works, of New York, manufacture latrines in this country, of plain, zinc coated, or enamelled iron. The bowl is formed in one piece with a section of the connecting pipe. The pipes are octagonal in section, and have flanges on the end by which they can be bolted together. The water is held at the same level in each of the bowls by means of a hollow trapped plunger. This plunger as well as the compartment in which it is placed is of the same kind as the one described in connection with Demarest's closet (Fig. 138). The same firm manufactures a number of trough-closets. In closets of this kind there is no bowl; but a trough long enough to receive the number of seats desired. The outlets to the trough are closed by a hollow trapped plunger according to Demarest's patent. The plunger operates in a chamber separated from the trough, and it is lifted by a combination of levers, which are clearly shown in the illustration. (Fig. 207.)

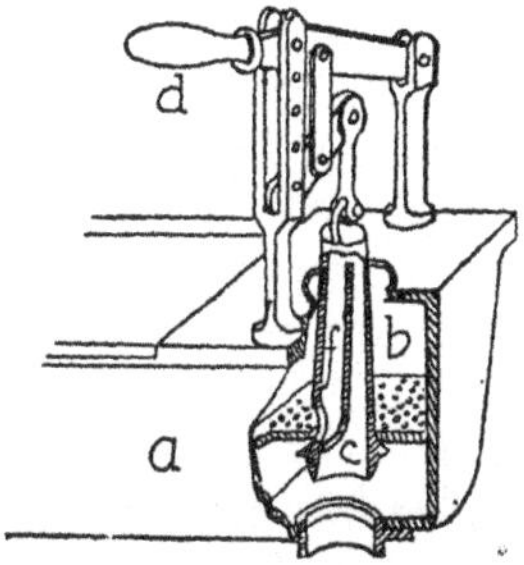

Fig. 207.—Trough-Closet.
a Trough. *b*, Plunger-chamber. *c* Plunger. *d*, Lever-handle. *f*, Overflow.

Parfitt's Latrines.—A. G. Myers, of New York, manufactures a latrine formed in the shape of a long wooden box lined with lead, the cover to this box being divided into as many seats as it is necessary to have. This box is connected by a six-inch outlet and trap to the main soil-pipe. The novel feature in this latrine is the arrangement of a tilting-tank above the end farthest from the outlet. The tank is connected with, and is to be filled from wash-basin or

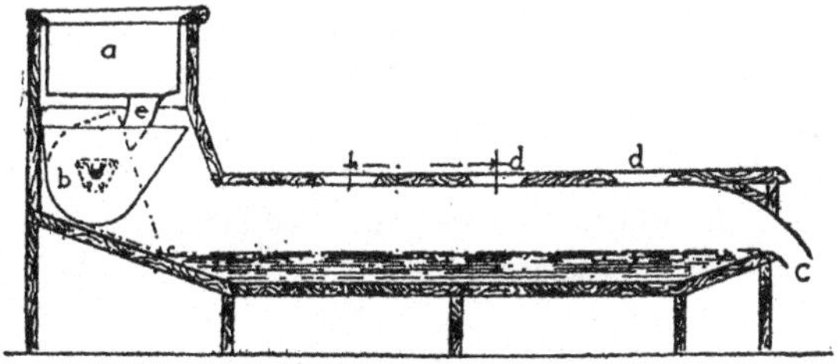

Fig. 208.—Parfitt's Latrine.
a, Sink or wash-bowl. *b*, Tilting-tank. *c*, Outlet. *d*, Seats. *f*, Lead lining.

slop-sink wastes. The tank being balanced on journals is so arranged that one end will be heavier than the other when it becomes filled with water. Tilting automatically it will empty its contents into the trough, and at the same time force the contents of the latrine through the trap.

In the place of latrines it is now a common, and I think the best practice, to use a row of long or short-hopper closets, either flushed by a tilting-tank, or a tank which is operated by connection with the seat or door of the water-closet room.

A tilting-tank can have its water-supply so arranged that the closet may be flushed at either short or long intervals. The time between the flushes should be regulated according to the abundance of the water-supply as well as according to the use the closet receives.

One or two arrangements for opening the supply-valve in the tank automatically will be described further on. Where rows of either long or short-hopper closets are used in place of a latrine, each bowl has the advantage of a separate and ventilated trap between it and the soil-pipe. While with the ordinary latrine we have each bowl with its connections to become foul, and these foul surfaces will be exposed to the room whenever the latrine is emptied. When foul matter is dropped into one bowl of a latrine, there it must remain, no matter how great or offensive the stench may be.

In hospitals it is often imperative to have the fæcal matter from patients discharged directly into the soil-pipe, in which case a detached closet, one of a row, would be found most serviceable.

I do not think any case could arise in which a row of hopper-closets would not be superior to a latrine unless it were in the case of a public park, where the closet would be open to the most common forms of public use and abuse.

SECTION VII.

SUPPLY-PIPE CONNECTIONS.

THE common manner a few years ago of joining the supply-pipe to the water-closet was by enlarging the mouth of the lead supply-pipe so it would fit against, or making it small enough to enter, the nozzle of the bowl; when the two were fitted, a piece of putty was put around it, and the whole covered with a piece of cloth held in place by a string. The joint is sometimes covered with a piece of sheet-rubber, the rubber being wrapped with wire instead of cord. Joints of this kind are rarely tight for any length of time, and they cause a great deal of trouble by their leakage. Years ago efforts were made to perfect the joint between the supply-pipe and the bowl; but the plumbers as a class do not appear to have taken advantage of the best methods.

As early as 1829, Tyler, of London, invented a nozzle which was to be bolted to the side of the bowl by means of a plate made in one piece with the nozzle, and shaped the same as a portion of the surface of the bowl. This plate had ridges, and the bowl corresponding grooves. When the plate was bolted to the bowl, and the space between the two was filled with red lead, or a cement of red and white lead, it made a good joint. The metal nozzle, being made of brass, either had threads to which a pipe could be screwed, or a lead pipe could be connected with it by means of a wiped solder-joint. Before making a solder-joint, the surface of the metal had to be made clean.

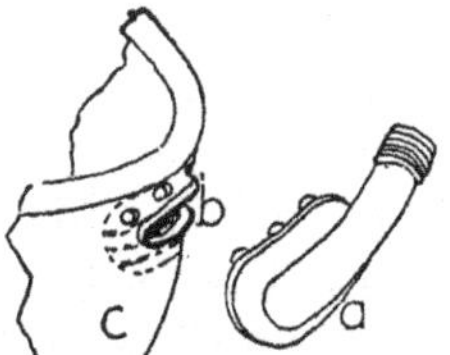

Fig. 209.
Tyler's Joint.
a, Brass nozzle and plate.
b, Opening into bowl with raised rim. *c*, Bowl.

In 1848, Armstrong invented a mode of connecting the supply-pipe with the closet-bowl. A metal ferule, with screw-threads on the outside edge, is imbedded in the nozzle of the bowl. Working on the thread is a clamp nut. A second brass ferule, with a flange around the larger end, is firmly screwed up against a rubber washer that rests on the first ferule. The lead supply-pipe can be connected by the usual wiped solder-joint to the brass nozzle.

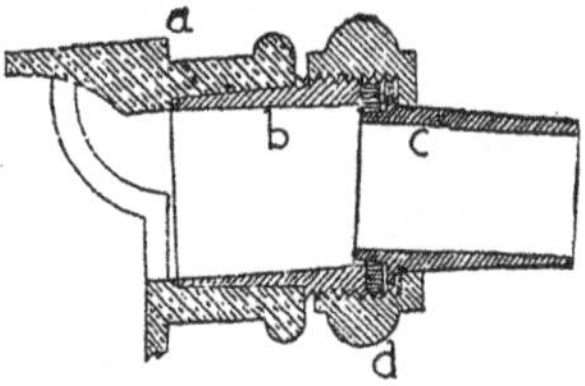

Fig. 210.
Armstrong's Connection.
a, Nozzle of bowl. *b*, First ferule.
c, Second ferule. *d*, Clamp nut.
e, Rubber washer.

A novel method of connecting the bowl with the water-supply pipe was invented in this country by W. S. Carr, of New York. A piece of brass tubing is bent or cast so as to form a right-angle with one end made smaller than the earthenware opening into which it is intended to fit. This metal tube has a flange or shoulder that fits against the outer edge of the earthenware nozzle, there being a rubber washer between the brass flange and the earthenware. Where the brass pipe bends, a hole is left through which a hook passes. This hook is made to fit around the earthenware where the orifice enters the bowl. Where the hook passes through the bend of the pipe there is a nut, by which the pipe can be screwed up tight against the earthenware nozzle. A lead pipe may be joined to this connecting pipe by a wiped solder-joint if it is brass, or it may be connected with iron or brass pipes by a screw-joint.

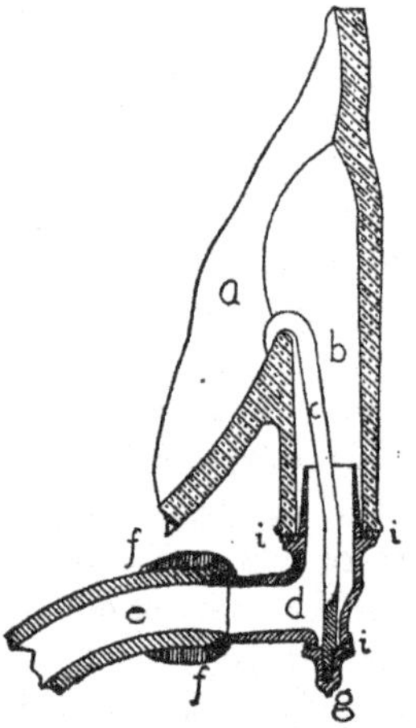

Fig. 211.—Carr's Connection.
a, Bowl. *b*, Earthenware nozzle.
c, Hook.
d, Brass connecting-pipe.
e, Lead pipe. *f*, Solder-joint.
g, Nut. *i*, Rubber washer.

I find patented in this country a manner of connecting the supply-pipe by putting it directly into the nozzle which projects from the bowl, when a piece of vulcanized-rubber tubing, which fits tightly around the pipe, is stretched over the earthenware nozzle. This manner of connecting the supply-pipe might be employed as a temporary expedient.

J. L. Mott has patented an ingenious mode of connecting the supply-pipe with the closet bowl. A small metal plate is used, which is zinc-coated or brass, and has a conical hole in the centre of it. The inlet formed on the bowl has a conical projection of earthenware that corresponds to the hole in the metal plate. A piece of lead pipe is fitted over the conical projection, when the plate is put over the pipe and screwed fast by means of clamp-screws introduced through slots which are made in the earthenware for the purpose.

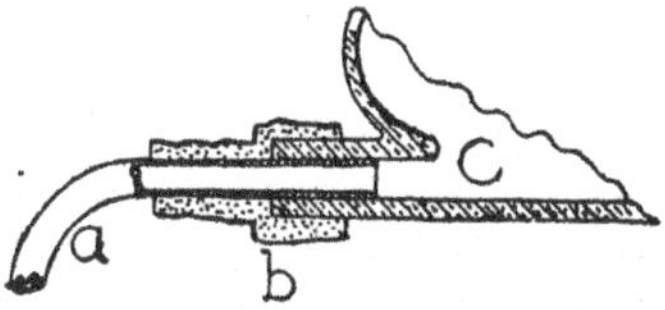

Fig. 212.
a, Lead pipe. *b*, Rubber-tubing. *c*, Bowl.

Soil-Pipe Connections.—With the most approved forms of water-closets where the trap is above the floor and forms a part of the closet instead of a part of the soil-pipe system, it becomes very important to have the joint between the water-closet and the branch from the soil-pipe made gas-tight; otherwise the trap and vent-pipes would be useless.

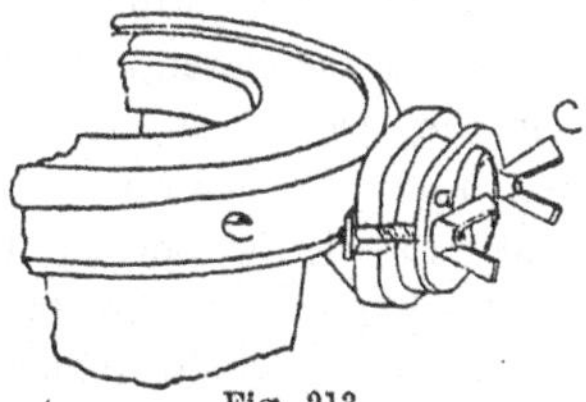

Fig. 213.
Mott's Supply-Pipe Connection.
a, Metal plate. *c*, Clamp-screws.
e, Bowl.

The common method of bringing the mouth of a trap, or a bell on the branch from the soil-pipe, and beating a part of the sheet-lead safe into it, then putting a little putty or cement on the floor, and screwing the closet down, is very defective. Where the sheet-lead is beaten into the bell it is a very imperfect joint, even when the space is filled with putty. The sheet-lead has no power to resist the packing of even

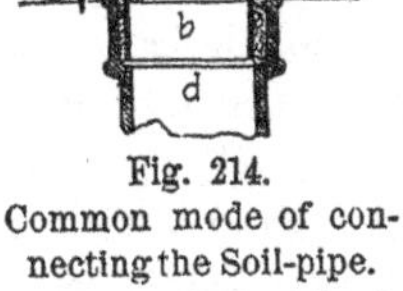

Fig. 214.
Common mode of connecting the Soil-pipe.
a, Outlet. *b*, Sheet-lead.
c, Putty. *d*, Soil-pipe.

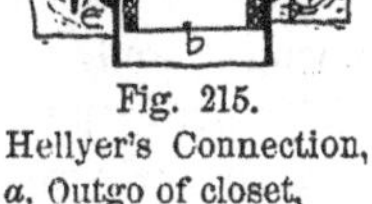

Fig. 215.
Hellyer's Connection.
a, Outgo of closet.
b Soil-pipe.
c, Cement.
i, India-rubber ring.

purty into the bell of the pipe. A joint of this kind might be opened, even when gas-tight in the first instance, by jarring on the floor, by any movement of the closet itself, or by the jarring of the soil-pipe.

Hellyer in his work describes a method of connecting a lead soil-pipe to the "Vortex" closet. A soft india-rubber ring is sent out with these closets, for fixing between the flange of the outgo and the "tafted" (bent over and beaten down) edge of the soil-pipe or lead safe. This india-rubber ring can be made to perfectly seal the jointing, and it will allow for considerable movement in the closet or the floor. The space between the flange of the closet which is screwed to the floor and the "tafted" edge of the soil-pipe or safe can be caulked with spun yarn and a cement composed of red and white lead.

Buchan in his work on plumbing illustrates his method of connecting the "Carmichael" closet to the soil-pipe. A lead branch is brought through a hole left in the floor for the purpose, and beaten out flat upon the floor by a dresser. The outlet of the closet, instead of having a flange, in the usual manner, has three grooves formed in the earthenware. A piece of lead is placed around the bottom of the closet outgo, and being forced into the grooves, and the depressions filled with solder, it would probably form a tight joint. The lead on the outlet of the closet and the tafted edge of the pipe are joined by solder.

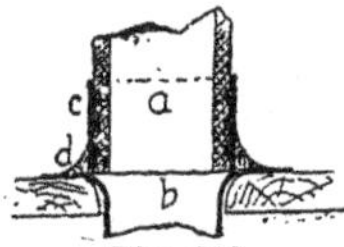

Fig. 216.

Buchan's Connection.

a, Outlet of closet.
b, Soil-pipe branch.
c, Lead around outlet.
d Solder-joint.

What is generally considered the best American practice is a modification of Hellyer's and the common methods. A Y-branch is brought from the soil-pipe to within a limited distance of the floor. Into the hub of the Y-branch a short section of brass pipe is caulked, by putting in a gaskin, then pouring in molten lead and driving it firmly in with a caulking tool. A short section of lead pipe is put in, just long enough to be carried through the hole made for the purpose in the floor, and beaten over so as to form a flange about two inches wide on the top of the floor, and at the same time reaching the brass pipe. When the lead pipe has been properly soiled and shaved, and the brass tube filed bright, they can be joined together by a wiped solder-joint. Where the pipe is turned over the floor, it is usually joined to the safe by a solder joint. There is a soft rubber ring stretched around the outgo of the closet just beneath the flange which is screwed to the floor. The space between the flange and the tafted edge of the lead pipe or the safe can be filled by a cement composed of white and red lead. This method of joining the closet to the soil-pipe is recommended by the most prominent sanitarians, but I think it objectionable in having the short lead pipe where it is exposed to the action of gases generated in the soil-pipe. There is every reason to believe that this acts on lead injuriously; in fact, reasons which are urged against lead soil-pipe may with equal force be urged against this short piece of lead; the whole inside sewerage system being only as efficient or as perfect as is its weakest point. I also object to having the closet depend for its stability on the floor, which is constantly undergoing expansion and contraction, according to the moisture and dryness of the air, and is subjected to vibrations from movements that may occur upon it. The above methods are all imperfect in their mode of connection. I think the methods described below are the best that have been introduced.

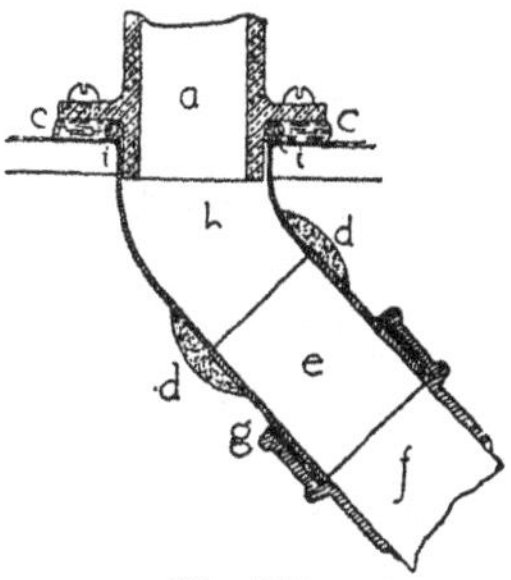

Fig. 217.
Improved Connection.
a, Outlet of closet. *b*, Lead pipe. *c*, Putty. *d*, Solder joint. *e*, Brass pipe. *f*, Iron pipe. *g*, Outlet-joint. *i*, India-rubber ring.

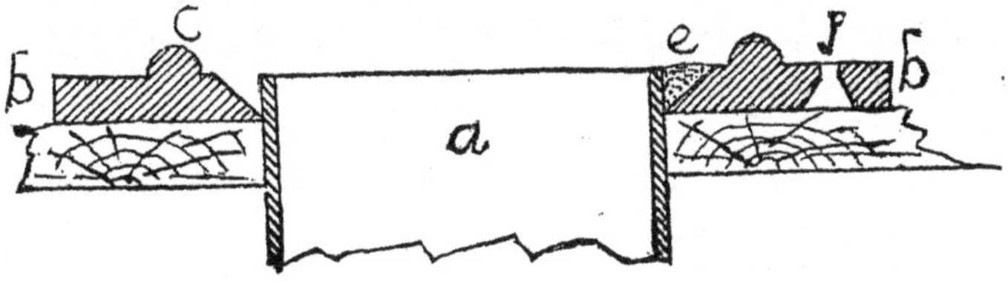

Fig. 218.

Fig. 218 represents the method adopted by the Meyer Sniffen Co. in connecting the Brighton closet with the soil pipe.

A brass plate that rests upon the seat is soldered to a short section of lead pipe, as shown in the cut. The flange on the outgo of the closet is bolted to this plate, being first bedded in red lead. There is a circular ridge on the brass plate that fits into a corresponding depression in the flange of the closet.

The Durham Drainage Company, of New York, have connected with the branch from the soil-pipe an iron fitting with a large plate or flange formed on it to receive the closet. When the outgo of the closet has a soft rubber ring put just beneath its flange, and it is screwed down tightly, a permanent gas-tight joint would be formed, the space between the two flanges which was not filled by the rubber ring being filled with cement. In section its appearance would be the same as Fig. 214, except an unyielding plate, which forms a part of the soil-pipe, would take the place of the yielding floor, and an iron pipe the place of the lead one. This method may be used where the trap, flange, and outlet are made in porcelain or earthenware.

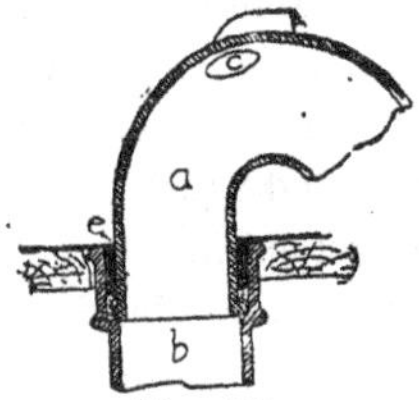

Fig. 219.
a, Trap. *b*, Branch from soil-pipe. *c*, Vent. *e*, Caulked joint.

The other method, which can only be used where the trap is of iron, has been recently introduced by J. L. Mott in connection with his side-outlet short-hopper closet. A branch from the soil-pipe has its hub brought to a level with the floor. The end of the trap is formed, as it is usual, with the spigot end of the soil-pipe with a small projecting ring around it. The trap is formed so it can run below the floor, the depth of the hub, into which it is caulked in the usual manner. In either of the cases mentioned, the trap of the closet and the closet itself forms a part of the soil-pipe system, the joints being as perfectly formed as any other joints in the system.

Safes.—It is customary and usually necessary to have sheet-lead safes under closets, to catch leakage or splashing that may occur from water or slops thrown into the bowl. These safes are simple shallow pans, formed by the turning of a piece of sheet-lead up about an inch-and-a-quarter high. These pans are placed beneath the closet, and their size varies, according to circumstances, from a safe which covers the whole of the bath-room floor to one that will only cover a space about two feet square. Safes, to be useful in case of an emergency, must have a waste-pipe. A few years ago (and the same manner of connection may be found in a large number of houses at the present day) it was the custom to connect this waste-pipe with the soil-pipe. In some cases there was not even a siphon-trap between the soil-pipe and the room. In the better class of work it was usual to connect the supply-valve with the trap of this waste by a small quarter-inch pipe, called a "weeping" pipe. This weeping-pipe was intended to carry the water that remained in the short section between the valve and the bowl when the water is cut off, into the trap of the safe-waste. It is positively wrong to connect the safe directly with soil-pipe. The weeping-pipe may be stopped without the knowledge of any one, leaving the water in the trap to evaporate. The trap of the waste would be siphoned by each discharge from the closet. The best practice is to carry a "tell-tale" pipe from the safe into the kitchen or cellar, and leaving it open over a sink that is in constant use, where if anything were wrong, it might be noticed and stopped. In England the custom is to carry this waste-pipe through the wall, where it is allowed to form a drip or overflow. In some instances the outlet is closed by a flap-valve, to be opened by the pressure of water, if the hinge has not in the meantime become rusty. North of the Potomac, in this country, our cold winters would prevent a waste-pipe of this kind from being effective, as a small leakage trickling through would be sure to freeze and stop up the waste-pipe.

Where the floors of bath-rooms are made of impervious materials, such as glazed tile or slate laid in cement, they should be made to drain to a convenient point, and have a waste-pipe similar to the lead safe.

Slop or Urinal Safes.—All prominent firms who manufacture plumbers' supplies in this country and England furnish these safes to fit in beneath the seat and over the bowl. These safes are

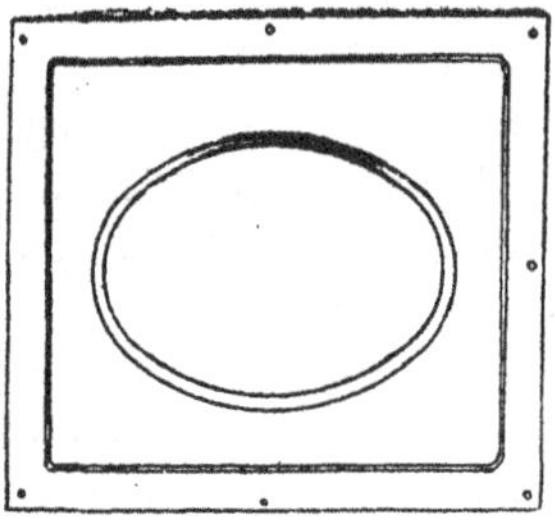

Fig. 220.—Oval Slop Safe.

simply a square, dished covering for the bowl, which has either a circular or oval hole about the size of the hole in the seat, with a surface of an impervious character formed so as to

Fig. 221.—Section.

conduct water or urine into the closet bowl. They are made from solid earthenware, glazed, or from iron or copper enamelled. In all cases they are useful and cleanly in fact and in appearance. Thin metal safes are the most convenient for fitting over closets in which the wood-work has already been put in position, but from the liability to chipping, which even the best enamel has, a glazed earthenware safe is to be preferred where a new closet is being fitted up. Where the closet is to be used as the urinal, a slop-safe is very important, as it prevents dripping on the wood-work or into the safe below the closet. In either case the rapid decay of urine would generate dangerous and offensive gases. A poor substitute is sometimes found for a slop-safe, by beating sheet-lead over a wooden piece shaped in a manner similar to the regular slop-safe. Although better than wood, a substitute of this kind is a mere make-shift, and should never be used in place of an enamelled or porcelain one.

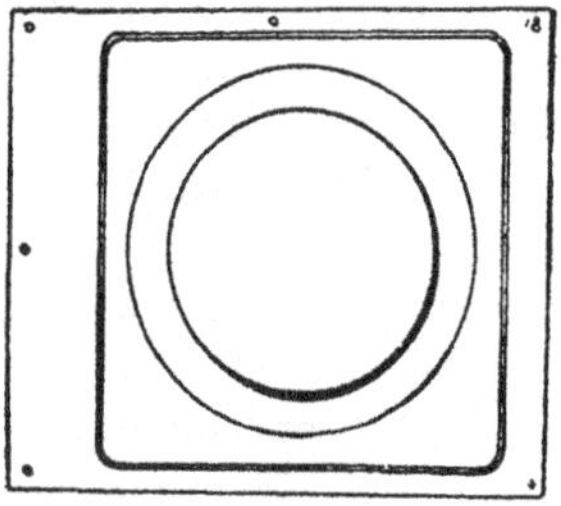

Fig. 222.—Circular Slop Safe.

WATER-CLOSET SEATS.

All are familiar with common boxed water-closet seats, often fine and expensive pieces of cabinet-work, fair without but foul within. These boxes secrete dust, dirt, and in many instances leakage and drippings from the closet. Seats of this kind, when made in their best form, have both seat and riser to open upon hinges, stout brass ones being the best: iron rusts easily, and the hinges become useless. When the riser and seat open the housekeeper can easily see that this space is kept clean.

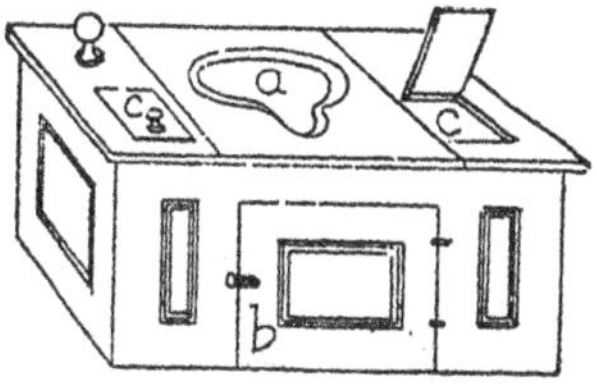

Fig. 223.—French Closet Seat.
a, Seat, *b*, Door in riser. *c*, Paper-box.

A convenient shape for the hole in a seat is egg-shaped, eight inches wide and ten inches long, and commencing about three inches and a half from the edge of the seat. This form of hole was patented in England thirty years ago. The manner of such work must be in keeping with the other wood-work in the bath-room. Heavy rubber duck is sometimes used instead of brass for hinges to the seat.

I take an illustration from Liger of a French panelled box water-closet seat. The riser has a panelled, hinged door. The hole in

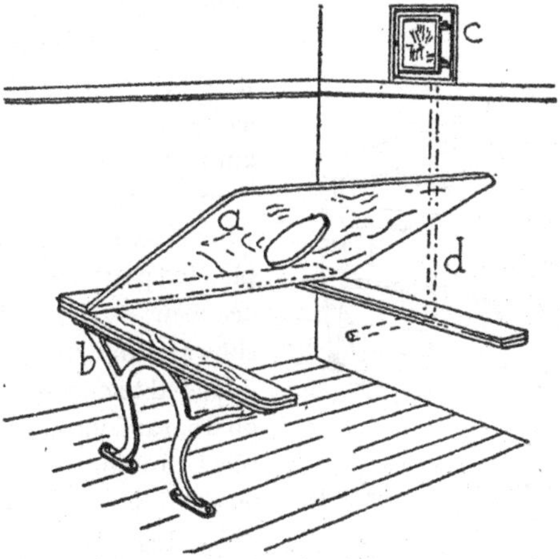

Fig. 224.—Open Seat.
a, Seat. *b*, Legs. *c*, Gas-jet for ventilation. *d*, Ventilating-pipe.

the seat is decidedly pear-shaped, being very much like the ones represented in the closet from Herculaneum, which must have been in use more than eighteen hundred years ago.

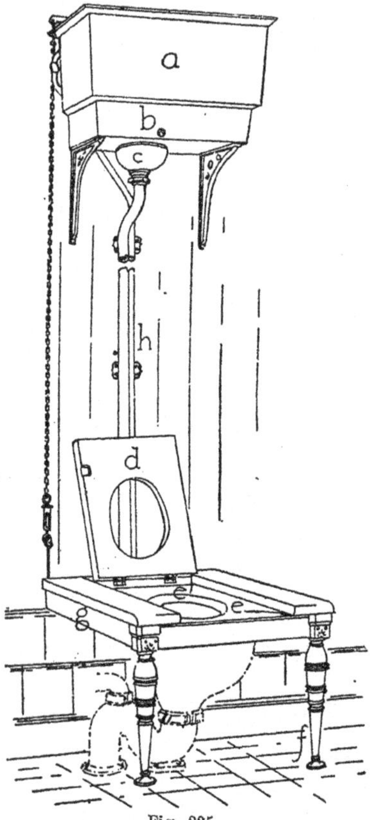

Fig. 225.

a, Tank. *b*, Service-box. *c*, Flushing-pipe. *d*, Seat. *e*, Slop-safe. *g*, Seat-frame. *h*, Supply.

I give here two illlustrations of what I consider the best forms of water-closet seat. The seat is supported by legs, which may be as ornamental as desired, and should be of some impervious material. The seat itself consists of a simple hinged leaf or board with a hole in it. A seat of this kind is furnished with the Brighton closet (Fig. 195). When not in use, the seat is turned back against the wall, exposing the white porcelain closet setting on a white or colored glazed tile floor, producing a neat appearance, and it is in fact very cleanly and easily cared for. A seat of this kind may be used with any form of closet except those which have unsightly or complicated parts that require concealment. The illustration shows a ventilating pipe that is intended to create a current of air from and through the closet bowl, carrying off any local offensive odors that may occur.

In Fig. 225 is shown a seat, the front of which is supported by legs, a slop-safe and tank in position. This arrangement as furnished by J. L. Mott & Co.

has either brass or nickel-plated legs, which are intended to rest upon a tile floor. Similar seats of this kind, having plain iron legs, are furnished by the different manufacturers. The slop-safe is let into the frame of the seat. The seat proper being thrown back against the wall when it is not in use, leaves the enamelled surface of the slop-safe exposed. This closet is flushed by the weight of the person using the closet. The seat presses down on a short lever, which communicates motion to the valve of the tank, by means of a lock chain. The floor and side of the bath-room are intended to be tiled. It is scarcely necessary to say that this makes a clean, healthy and efficient apparatus for the purpose it is intended to fulfil.

For prison use, or in places where the same abuse is likely to occur, it is best to use a jump-up seat like the one illustrated in Figure 226. The seat is oval, formed of wood, and is just large

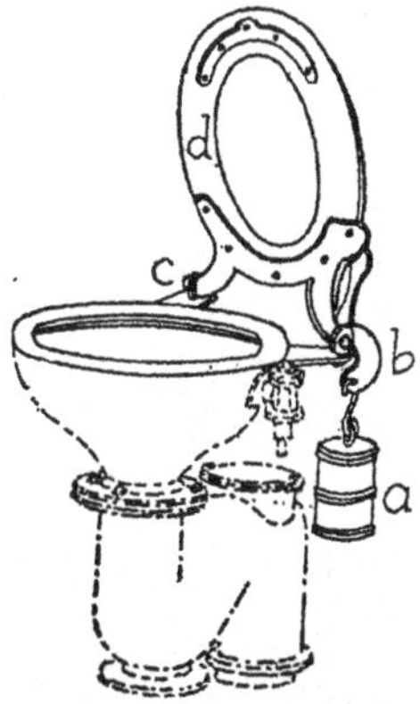

Fig. 226.—Jump-up Seat
a, Weight. *b*, Lever.
c, Journals. *d*, Seat.

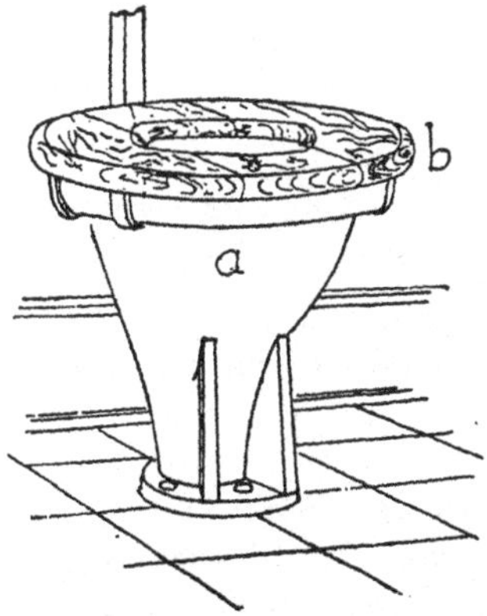

Fig. 227.
a, Closet. *b*, Seat. *c*, Clamp.

enough to cover the bowl of the closet. Two iron journals are screwed, by means of an iron plate from which they project, to the wooden seat. One side has a lever and weight attached to the journal. By means of the weight the seat is kept thrown back against the wall, unless it is held down over the bowl by a greater weight.

Sometimes the long hopper porcelain closet is set on a tile floor and has a wooden seat which is held in position by iron clamps, that are clamped around the earthenware rim.

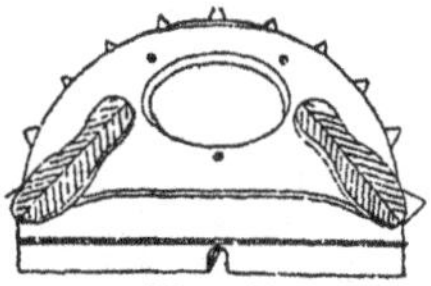

Fig. 228.
Seat for Standing on.

Liger describes a closet seat or stand which is intended for use in the crouching or Oriental fashion, instead of as a seat. Places are arranged for the feet, being made rough or ridged, so there will be no danger of slipping.

Levers.—It seems advisable to describe the two following simple levers for opening the valve in the flushing-tank by pressure on the seat:

Fig. 229 is a double lever which has its pivots or fulcrums attached to the under side of the seat. Motion is imparted to the first lever by pressure on the seat, while it in turn moves the short arm of the second lever. The long arm of the second lever is connected by a chain with the lever that operates the flushing valve in the tank.

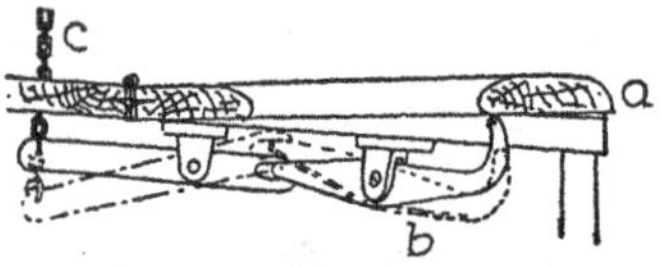

Fig. 229.—Lever for opening Supply-tank.
a, Seat. *b*, Lever. *c*, Connecting-chain.

The arrangement used by Henry Huber & Co., of New York, in connection with his "Tidal Wave" closet is simple, and not liable to get out of order. It is a simple lever which moves up or down on two journals, the bearings in which they work being screwed to the floor. The lever is pressed down by rods which are attached to it; these run through guides and come in contact with the under side of the seat. The weight of the tank lever keeps the seat slightly raised

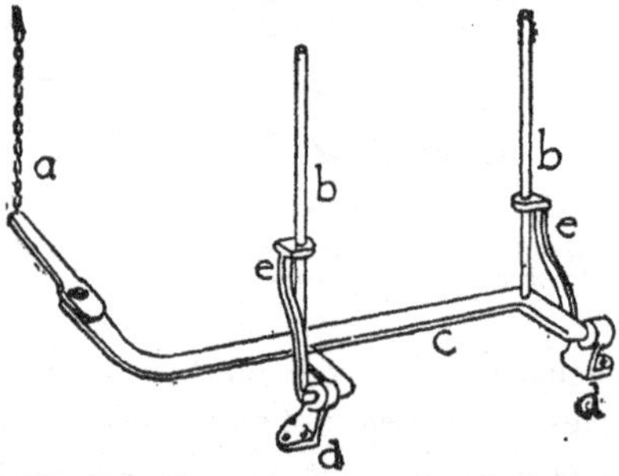

Fig. 230.—Lever for opening Supply-Tank.
a, Chain. *b*, Rods. *c*, Lever. *d*, Journal boxes. *e*, Guides.

unless there is sufficient weight on the seat to counteract its effect.

Bidet Attachment.—Cragie's bidet attachment is a small branch from the supply that is attached to the inside of the water-closet bowl. When not in use it is turned up under the seat or urinal safe; when in use, it is turned down into the centre of the bowl. From it a small jet or spray of water is thrown up toward the seat. This attachment is important in some diseases.

Quite a number of devices have been invented for the purpose of disinfecting water-closets. These "germicides," as some of them are called, are intended to destroy germs and odors that may be generated in receivers or other concealed compartments that may be connected with badly designed water-closets. Instead of using disinfecting apparatus and a bad or poorly constructed closet, it is far preferable, both on the score of health and comfort, to select a simple, cleanly closet, with its trap properly connected with the soil-pipe and ventilated, and thus have no need of a "germicide."

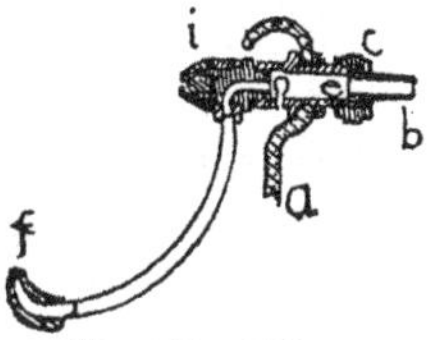

Fig. 231.—Bidet.
a. Bowl. *b*, Supply.
c Nut. *e*, Valve.
f, Nozzle for jet.
i, Packing.

SECTION VIII.

SUPPLY VALVES AND TANKS.

WATER-CLOSETS are supplied or flushed with water by means of supply valves and tanks, or cisterns.

Supply valves may be automatic in their operation, or opened by the party using the closet.

The simplest form of supply valves are ground and compression cocks, which are opened respectively by a lever and screw. Between these simple valves and the complicated diaphragm valves there are many deviations.

The following valves which I will describe are the ones that have been most extensively used in connection with water-closets.

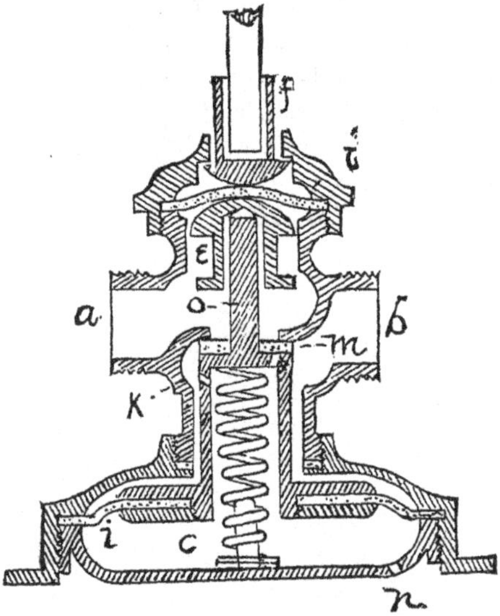

Fig. 232.—Bartholomew's Supply-Valve.
a, Inlet. *b*, Outlet. *c* Air chamber. *o*, *f*, Valve rods. *i*. Diaphragm. *k*. Water inlet to air chamber. *m*. Washer. *n*, Leak or drip.

Bartholomew's Supply-Valve.—This valve has been extensively used in connection with the pan-closet of the same name (Fig. 120). This supply valve was intended to close itself after a certain amount of water had passed through it into the closet. A projecting disc into which a diaphragm is fitted forms the bottom of the valve. When the valve is first opened, either by pressure upon the seat or by means of a lever, the chamber above the diaphragm would be immediately filled with water, and it would be impossible for the valve to close until the air chamber beneath the diaphragm fills with water through a small hole made for the purpose.

The t'me the valve occupies in closing is governed by the size of this orifice. A spiral spring forces the valve back against its seat as soon as the pressure above and below the valve has become equal. The water in the bottom compartment wastes through a small hole, and leaves the valve ready again for action. Fig. 233 is a later pattern of Bartholomew's supply-valve. The spring in this one is omitted, and gravity is depended on to close the valve when the pressure in the two compartments has been equalized.

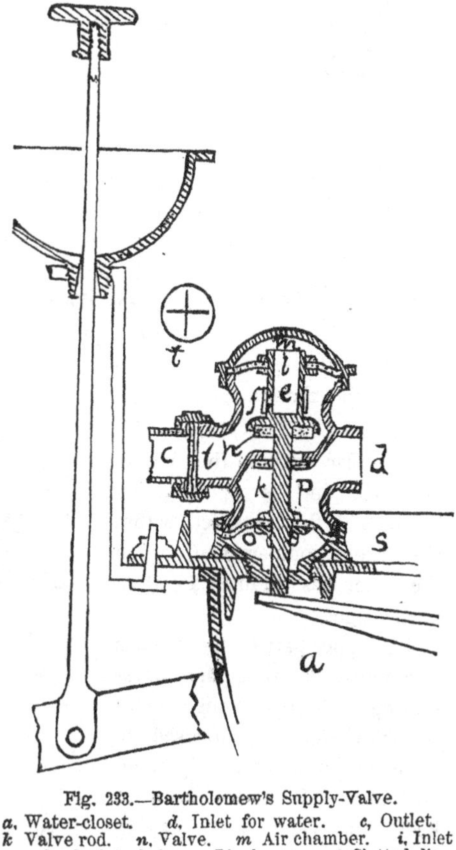

Fig. 233.—Bartholomew's Supply-Valve.

a, Water-closet. *d*, Inlet for water. *c*, Outlet. *k* Valve rod. *n*, Valve. *m* Air chamber. *i*, Inlet and *e* Outlet of air. *o* Diaphragms. *t*, Slotted disc. *s*, Drip pan. *k*, Rubber valve. *p*, Washer.

The air chamber is situated in the top of the valve, and the water passes into it through a hole left for the purpose, and leaks out through a small hole, which is covered by a rubber valve that opens only in an outward direction. To assist the valve by retarding the flow of water, so it will fill the upper chamber, a slotted disc is inserted in the outlet of the valve.

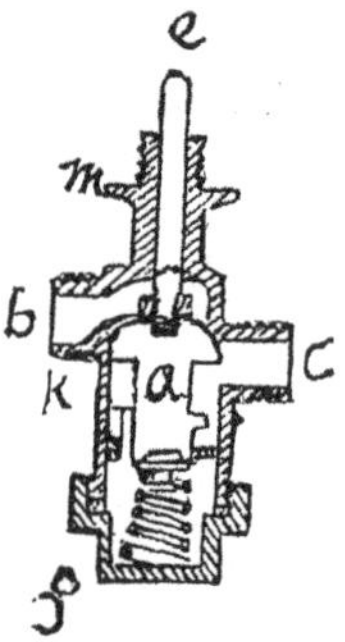

Fig. 234.
a, Valve. *b*, Inlet. *c*, Outlet. *e*, Rod. *f*, Spring.

The Climax Valve.—The valve used in connection with Demarest's Climax Closet is simple. It consists of a piston-shaped valve, which is held in position, or against its seat, by means of a spiral spring. The valve works between guides. Screwed into the top of the valve is a small rod, which, when pressed down, opens the valve. When the rod is relieved from pressure, the valve is closed by means of the spring in the bottom of the chamber.

Demarest Closet-Valve.—This valve is placed within the plunger chamber, and is operated by means of a float situated within the same compartment, when the water has attained a certain fixed height.

The upper part of the valve is perforated, so that the water passing through the same sprinkles, and is supposed to wash the plunger and plunger chamber. It seems hard to keep this valve from leaking in practice. A spring in the bottom of the valve assists in keeping the valve against its seat.

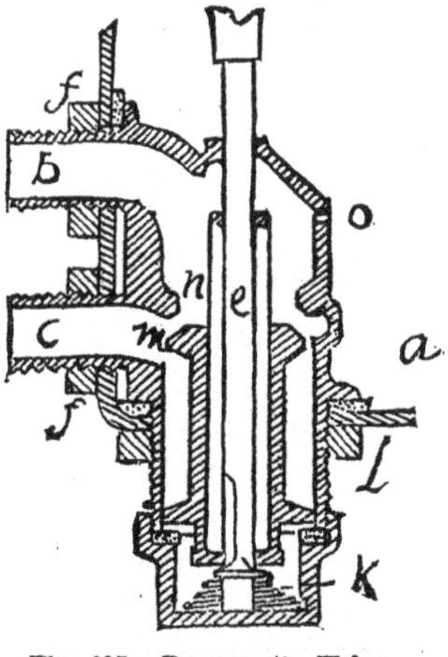

Fig. 235.—Demarest's Valve.

a. Plunger chamber in closet. *b*, Outlet. *c* Inlet. *e*. Rod connecting with float. *m*. Valve. *n* Valve seat. *f* Nuts to attach to closet. *k* Spring. *o*. Perforation for sprinkling plunger chamber.

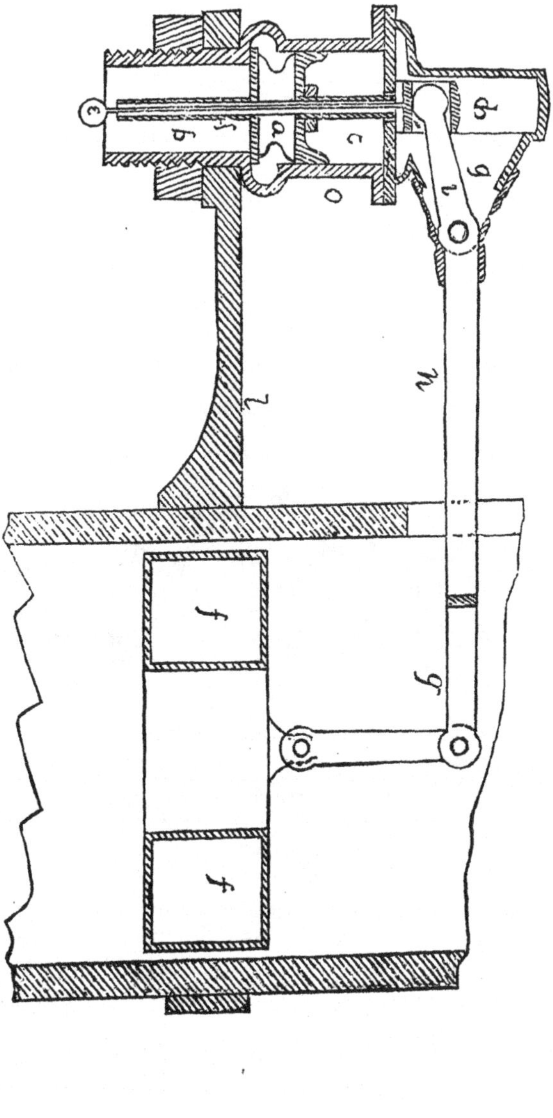

Fig. 236.—Jennings' Hydrostatic Valve.

a, Diaphragm. *b*. Water inlet. *c*. Compartment above diaphragm. *d*, Lever compartment.
e, Ball valve to close inlet to the diaphragm. *f*. Float. *g*, Lever. *l*, Bat connecting valve with closet.

Jennings' Hydrostatic Valve.—This valve is also operated by means of a float which encircles the plunger (see Fig. 130). When the water-closet bowl is emptied by raising the plunger, the float drops, the valve is opened, and the closet is flushed. When the float rises it opens an orifice, which lets the water into a diaphragm chamber. The pressure of the water in this chamber closes the valve gradually. The different parts of the hydrostatic valve have to be so nicely adjusted to themselves, and to suit the pressure of the water supply, that they are usually troublesome, and it is hard to keep them from leaking.

Kelly's Stop and Waste Cock.—This cock or valve is manufactured in Chicago, and it will be readily understood by reference to the illustration (Fig. 237).

Two plugs or valves are attached to a crank that moves in a

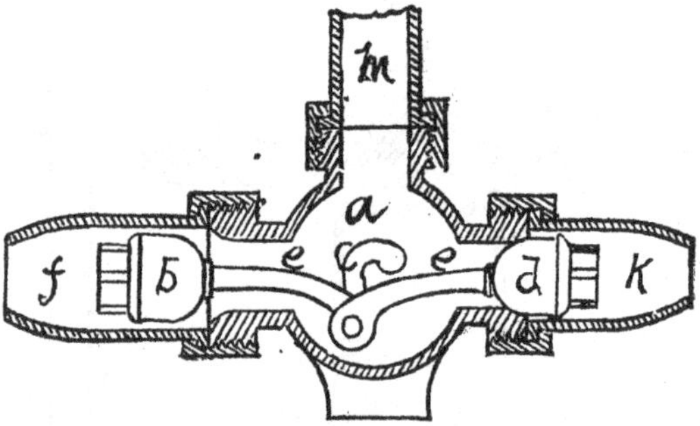

Fig. 237.—Kelly's Valve.
a, Inside of valve. *b*, Plug to inlet. *d*, Plug to outlet or waste. *c*, Crank that opens and closes plugs. *e*, Connecting arm. *f*, Inlet for supply. *k*, Waste. *m*, Opening to the flushing outlet.

centre spherical chamber. When the inlet plug is open the waste plug is closed, and the water supply will flow into the closet. Where it is necessary to place a closet in an exposed situation, this supply valve can be used and placed beneath the frost line, when the water in the pipe between the valve and the outlet in the closet would run to waste through the waste valve, and leave no water in the pipe to freeze.

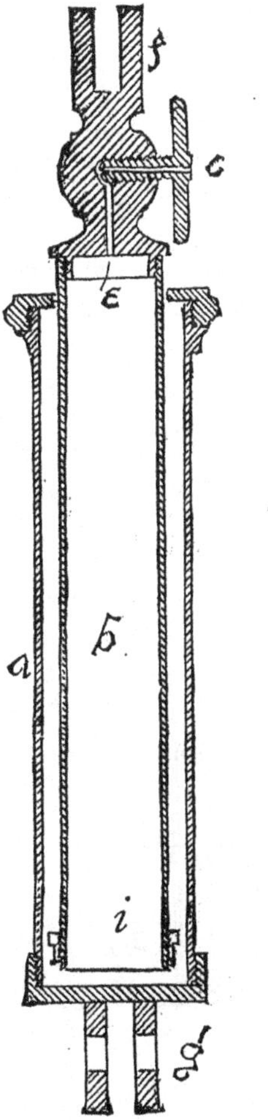

Fig. 238.
Underhay's Regulator.
a, Large cylinder. b, Small cylinder. e, Air hole.
c, Thumb-screw with air hole in it. f, Point of attachment to valve lever.
g, Point of attachment to the floor. i, Cupped washer.

Underhay's Waste Preventer.—In connection with supply valves I note and illustrate Underhay's Waste Preventer, designed to close supply valves when a certain amount of water has passed through the valve. This regulator consists of two cylindrical chambers. The small one, which is closed at the top, moves up and down freely in the large one that is closed at the bottom.

The small cylinder has on its lower end a cupped leather piston, and on its upper end an air hole; the size of which is regulated by a thumb-screw. The lever that opens and closes the valve is fastened to the regulator by a bolt and slot in the top of the small cylinder. It can be readily raised, as the cupped leather acts as a valve, and lets the air enter the small cylinder freely. When the small cylinder commences to descend by the action of gravity, the air in it would be compressed, and could escape only through the air hole in the top. The leather cup would be pressed tightly against the sides of the large cylinder. The cup dips into a lubricating oil, and in its passage up and down lubricates the side of the large cylinder. The regulator is illustrated in position in Figure 72.

REMARKS.

Water-closet supply valves that are connected directly with the supply pipe of the building must be condemned, from a sanitary standpoint, because there is a possibility, if not a probability, of the water supply becoming contaminated by suction of either dirty air or water from the bowl of the closet. This may take place when a valve or faucet is opened on a floor below the closet, at the same moment when the water-closet valve is open, the water or air passing back from the closet into the supply pipe and out through the open faucet below.

Valves must also be condemned on the score of economy, as it does not seem possible to prevent them from leaking and keeping up a constant waste of water, and repairs are always needed.

Their principal defect as a water-closet supply is that they do not let the water out at once in a large quantity so as to cleanse the closet thoroughly. Where a large flush can be emptied directly into the bowl it will have the desired cleansing effect, while triple the quantity would be of no avail if supplied in dribblets.

WATER-CLOSET TANKS.

Tanks have been used at least a hundred years for the purpose of storing water, so as to flush or wash water-closets.

Both the Cummings' and Bramah closets, two of the earliest

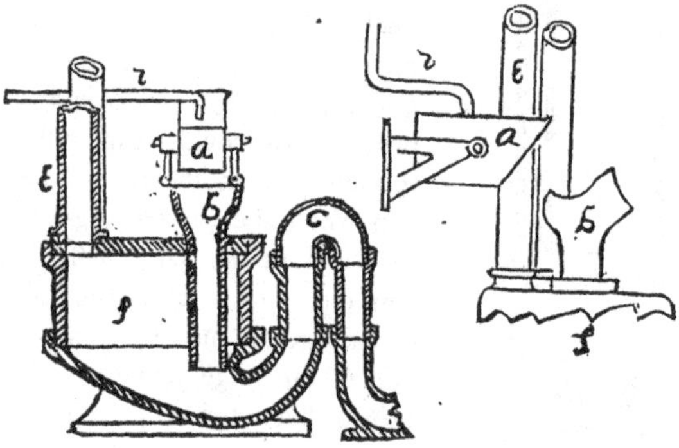

Fig. 239.—Shones' Tank.

a, Tumbling tank. *b*, Funnel-shaped inlet. *c*, Syphon. *f*, Reservoir. *i*, Supply pipe. *e*, Waste pipe for utilizing waste water for flushing purposes.

English closets, and patented before the nineteenth century, were flushed by means of tanks.

It has not been many years since tanks were discarded in favor of some of the many patented supply valves.

Tanks for flushing closets are fast coming into general use again as their utility is appreciated. Tanks may be treated under the head of automatic and non-automatic.

Shones' Syphon Tank.—This tank is simply a reservoir with an outlet through a syphon, and a funnel-shaped inlet at the top. The reservoir is filled by means of a small tumbling tank. The last time the small tank empties, when the reservoir is nearly full, the syphon will be started, and the whole contents emptied into the closet.

McFarland's Tumbling Tank is simply a small receiver or bucket hung on journals. The method of its operation is clearly shown in the cut. By regulating the flow of water from the supply cock the tank can be made to empty at long or short intervals.

Either of the last-mentioned tanks are simple in their construction, and, if the journals and journal boxes are well made, they will not be liable to get out of order, and will answer well for use in

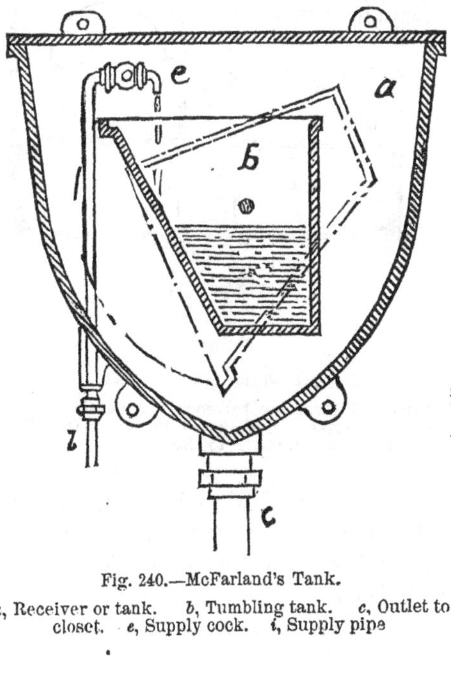

Fig. 240.—McFarland's Tank.

a, Receiver or tank. *b*, Tumbling tank. *c*, Outlet to closet. *e*, Supply cock. *i*, Supply pipe

public places where persons using the closet would be too careless to operate a non-automatic tank.

Among non-automatic tanks I will describe a small number of plug and syphon tanks.

Tylor & Son's Plug Tank.—This tank, which is illustrated in Fig. 241, seems a fair example of its class. It has a tank proper and a service box beneath. A plug which is held in position by a weighted lever stops the outgo from the service box. When the lever is pulled down a plug attached to the same lever on the opposite side of the fulcrum closes the communication between the tank and the service box. Unless the plug between the tank and service box fails to take its seat properly, only the water in the service box can be discharged at one time into the closet, although the lever may be held up any length of time. This tank as well as nearly all non-automatic tanks, are supplied with water through a ball cock. A floating copper ball by means of a lever shuts the water off when it has reached a certain height in the tank. This height is generally near the height of the outlet into the overflow. The tank is manufactured with a cover, as is usual in England.

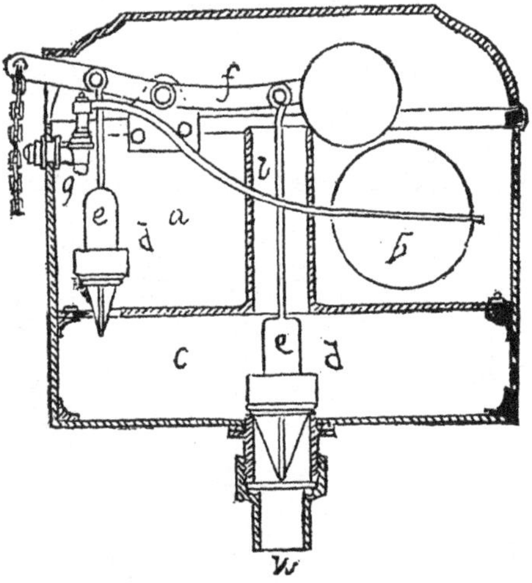

Fig. 241.—Tylor & Son's Tank.

a, Tank. *b*. Float for ball-cock. *c*, Service box. *e*, Plugs or valves. *f*, Weighted lever. *i*, Overflow. *g*, Supply valve. *n*, Outlet to closet.

The Duplex Tank, manufactured by J. L. Mott & Co., is a good example of a before and after-wash tank. This tank has three compartments, so arranged that while the service box is being filled the same pull of the lever opens the compartment that gives the

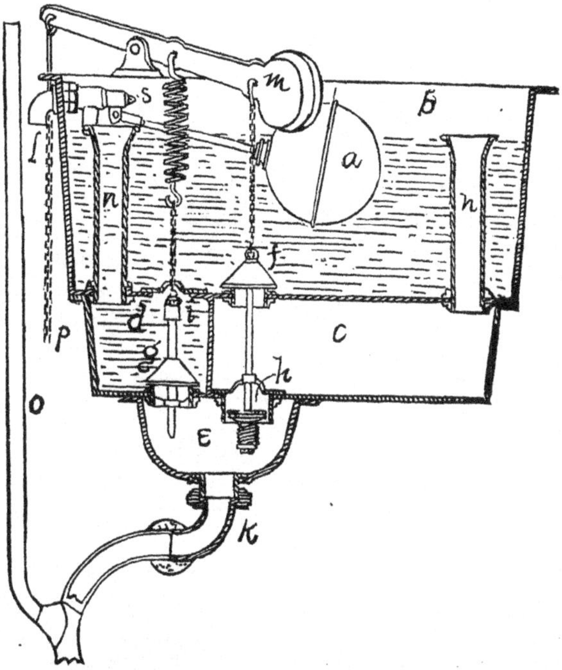

Fig. 242.—The Duplex Tank.

a, Float valve. *b*, Tank. *c*, Service box. *d*, Before-wash. *e*, Funnel-shaped outlet. *f*, Valve between tank and service box. *g*, Valve of before-wash. *h*, Valve to close service box while filling. *i*, Valve to close after-wash while emptying. *k*, Outlet to closet. *l*, Lever chain. *m*, Weighted lever. *n*, Overflow. *o*, Air pipe.

before-wash. As soon as the lever is relieved from weight, it takes its normal condition by the act of gravity, and opens the service box which flushes the closet.

The before-wash is of importance, as it prevents fæcal matter from adhering to the bowl. The valves or plugs have all either

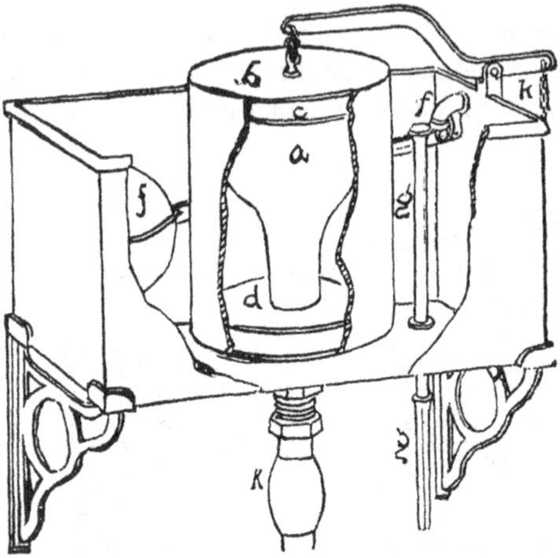

Fig. 243.—Doulton's Syphon Tank.

a, Syphon. *b*, Cylinder. *c*, Mouth of Syphon. *d*, Plate nearly the same diameter as cylinder. *f*, Float valve. *g*, Overflow. *h*, Lever. *k*, Outlet.

rubber washers or seats, so that they can easily make a water-tight joint.

An air pipe is shown attached to the outlet pipe below the closet. This prevents air being sucked through the water in the tank, and so prevents noise that would otherwise occur.

The Doulton Syphon Tank.—This tank is waste preventing, and its contents are emptied by means of a syphon. An annular syphon is surrounded by a cylinder, air tight on top and connected with the lever, that is operated either by the movement of the seat or by hand. To start the waste preventer the lever must be pulled down; this raises the cylinder and empties enough water into the syphon to start it into action. A two-gallon flush is obtained in this way without the necessity of holding the lever down for a moment.

The supply is governed by the usual float-valve, and there is also an overflow in case the supply valve should leak.

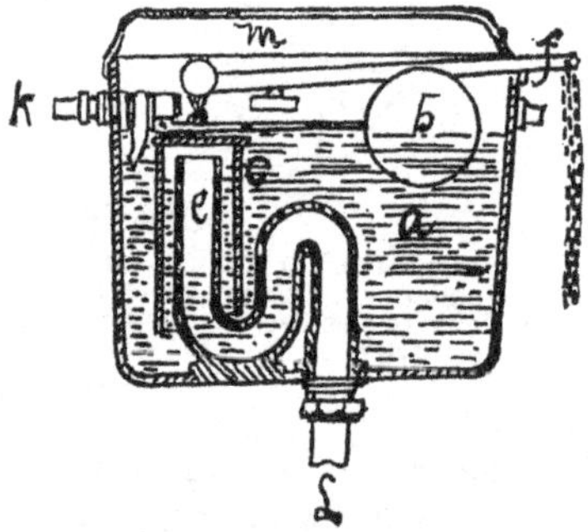

Fig. 244.—Winn's Syphon Tank.

a, Tank. *b*, Float. *c*, Cylinder. *e* Syphon. *f*, Lever. *k*, Supply. *l*, Outlet. *m*, Cover.

Winn's Syphon Tank.—This tank is manufactured in Birmingham, England. In this tank there is also a cylinder, which, when raised by the lever, empties water into and starts a syphon. When started the syphon would extract the water. The amount of water would be regulated by the size of the tank and the length of the cylinder. The illustration shows the tank covered, a custom almost universal in England, but not usual in this country. It serves a good purpose by preventing dust and other impurities from settling into the tank. Either of the two last-described tanks are simple, and have no valves (except the ball-cock) to get out of order. For this reason they are to be commended.

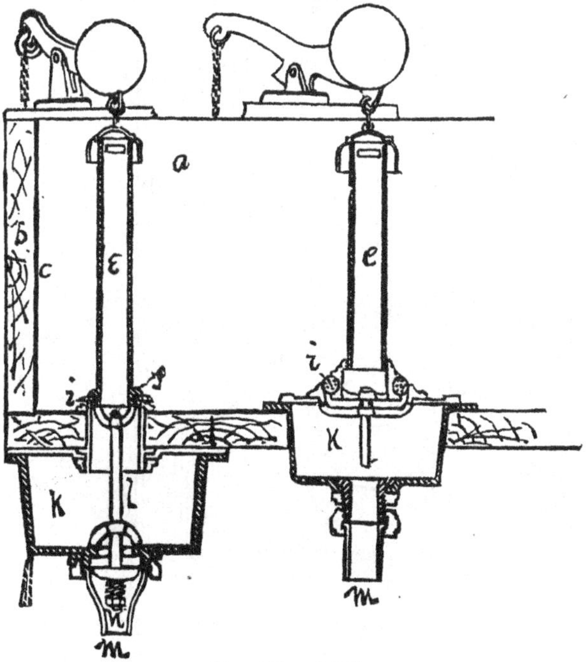

Figs. 245 and 246.

a, Tank. *e*, Valve and overflow. *c*, Lead or copper lining. *i*, Rubber flange. *k*, Service box. *l*, Connecting rod between two valves. *m*, Outlet. *n*, Valve to close service box while filling.

Figs. 245 and 246 show a method of using one large tank for a row of closets. The tank can be lined with either copper or sheet lead. The valve or plug is a hollow tube, being open near the top and acting as an overflow. At the bottom it is encircled by a rub-

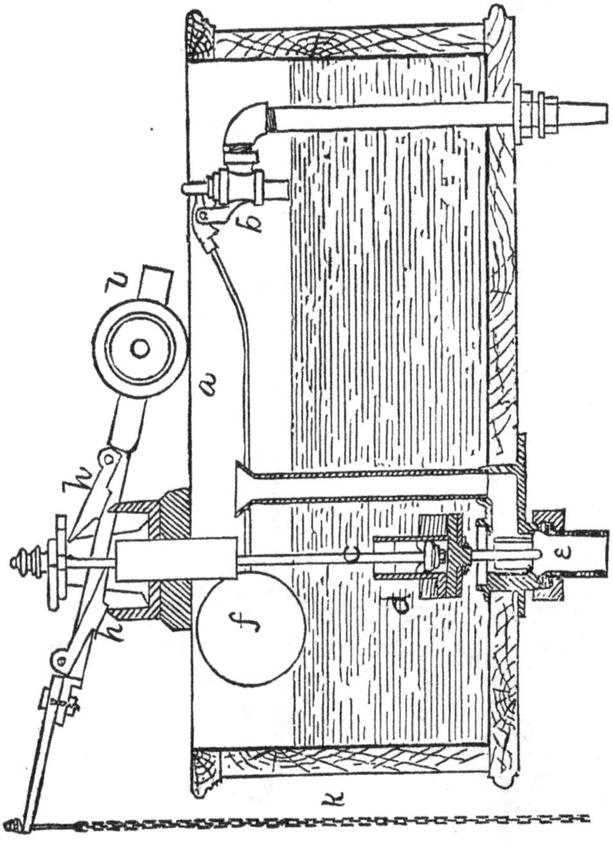

Fig. 247.—Boyle's Tank.
a, Tank. *b*, Ball-cock. *c*, Weighted piston-rod. *d*, Valve or plug. *e*, Outlet. *f*, Float. *i*, Weighted lever. *h*, Dogs or short levers. *r*, Chain connecting with hand-pull or seat.

ber flange or ring. This rubber rests on a seat prepared in the bottom of the tank.

Fig. 245 has a service box beneath the outlet, so arranged that when the plug is raised the outlet to the service box is closed by a second valve, but as soon as the lever is dropped the opening between the tank and service box is closed, the outlet of the service box is opened, and its contents discharged into the closet. Fig. 246 shows a non-waste-preventing plug.

Boyle's Tank.—The plug or valve is the distinguishing feature in this tank. A number of plugs may be used in one tank, each connected with a separate closet. By the arrangement of this plug a before and after-wash is obtained, and the cistern is made waste preventing if the mechanism is in proper order.

The lever, which moves on two trunnions in a journal plate, has two dogs or short levers pivoted in a slot of the large lever equidistant from the journals. These dogs are pointed in opposite directions and each rests on independent fulcrums.

In order to obtain the first wash, when the lever is pulled down the dog strikes against a projecting plate on the piston rod and raises the plug, and allows the water to escape and flush the closet. The dog which raises the piston has its fulcrum nearest the end of the lever where the motive power is applied. The valve, which slides up and down on the piston rod, takes its seat by the action of gravity. As soon as the hand-pull is released, the weight on the opposite end of the lever causes the plug to be again raised by the second dog. The position of the fulcrum of this dog causes the plug to be raised higher than it was in the first instance. By the arrangement of these dogs the desired amount of water may be obtained. The plug is a short piece of tubing with a leather or rubber disc on its bottom, so the plug can make a water-tight joint with the seat. In the tube is a piston with a soft leather washer. On the top of the piston rod is a weight, which forces the piston down to the bottom of the

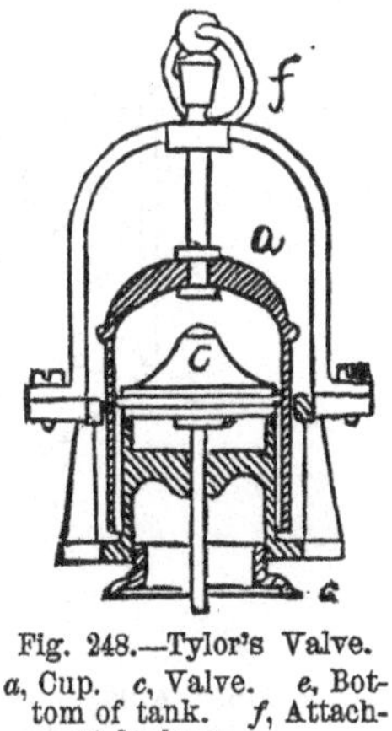

Fig. 248.—Tylor's Valve. *a*, Cup. *c*, Valve. *e*, Bottom of tank. *f*, Attachment for lever.

tube as soon as the long lever is horizontal, the water escaping past the soft leather cup. When the dogs or short levers raise the piston, the plug is lifted at the same time by capillary attraction.

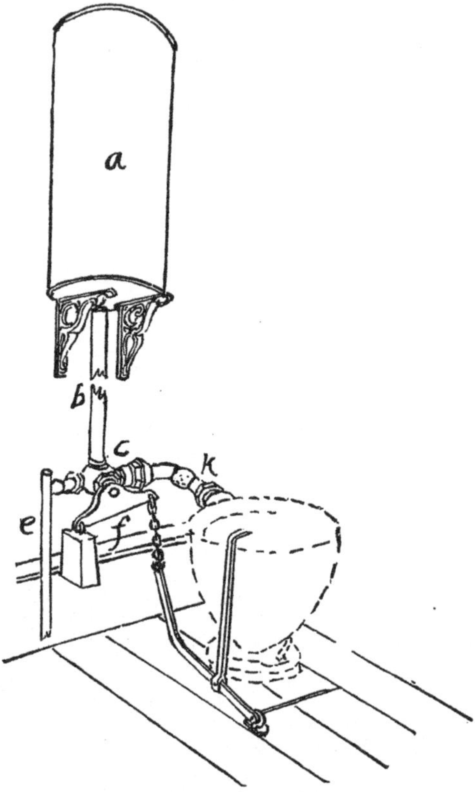

Fig. 249.—Kelly's Tank.

a, Tank. *b*, Supply-pipe. *c*, Valve (Fig. 237). *f*, Weighted lever. *k*, Water-closet connection.

Tylor's Waste-Preventing Plug.—Fig. 248 represents a waste-preventing plug to be used in connection with a tank, manufactured by Tylor & Son, of London. The valve in this case is within an

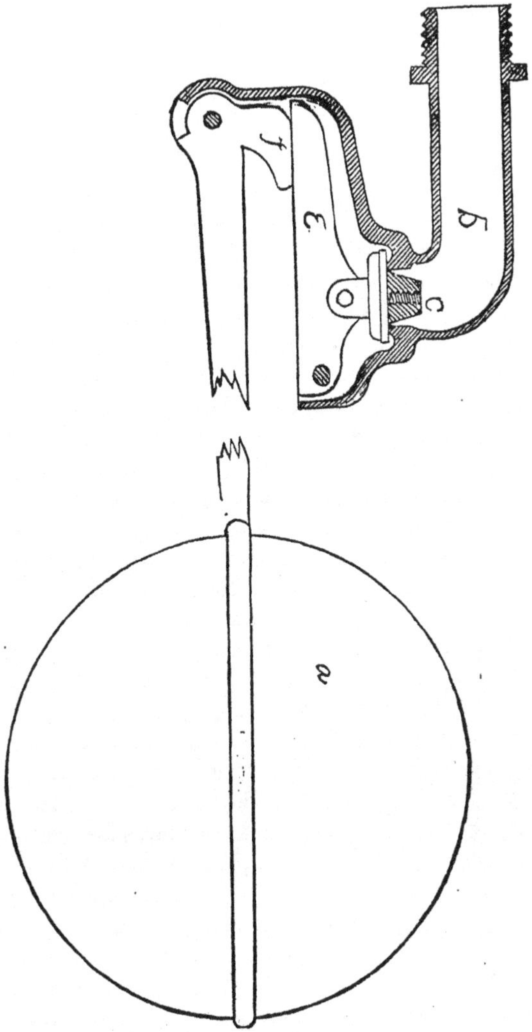

Fig. 250.—Tylor's Ball-Cock

a, Copper ball. *b*, Supply. *c*, Valve. *e* and *f*, Levers.

inverted cup. When the cup is raised the plug or valve is raised at the same time. If the cup should be held up for any length of time the plug would descend by the action of gravity, as soon as equilibrium was established between the upper part of the cup and the tank. This valve is generally regulated so the tank will discharge about two gallons of water each time the valve is raised.

Kelly's Tank.—In connection with his supply valve, Thomas Kelly, of Chicago, has a tank which is operated by compressed air. This apparatus also professes to be waste preventing. Weight on the water-closet seat by means of a weighted lever opens one side of the valve, when water will rush into the cylindrical and air-tight chamber above the closet. The air would be compressed according to the head of water in the street mains. As soon as the water-closet seat is freed from the imposed weight, the weighted lever would open the outlet side of the valve, when the water under pressure would be driven through the flushing-rim of the closet (see Figs. 237 and 249).

As almost without an exception tanks are supplied with water through ball-cocks, I have illustrated some of the best in the following cuts.

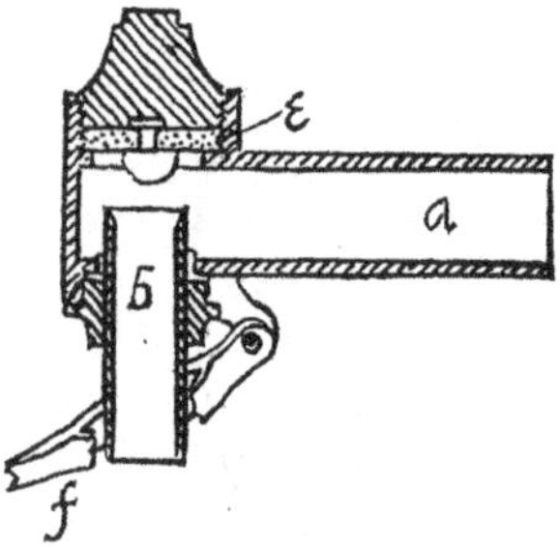

Fig. 251.

The balls are always of spun copper, and, being hollow or filled with air, they float. The ingenuity of inventors has shown itself in the methods adopted in closing the supply pipe with the lever, which is moved by the floating ball. The illustrations scarcely need description as they explain themselves.

Fig. 250 is one of Tylor & Son's, of London. The valve is forced against its seat by a double lever. This is sinple and effective.

Figs. 251 and 252 are taken from Bailey Denton's work. One

is a simple tube, which, when closed, presses against a rubber seating. The other is a valve with a piston and piston rod attached, which aids in keeping the valve against its seat.

Tanks should be connected with the closet by at least an inch-and-a-quarter pipe. The bore of the outlet of the tank, the inside diameter of the pipe, or the inlet to closet should never be diminished in any way. Plumbers, in some cases, diminish the opening by beating the lead pipe together, in bending it, by washers, red lead, and putty used in making the couplings.

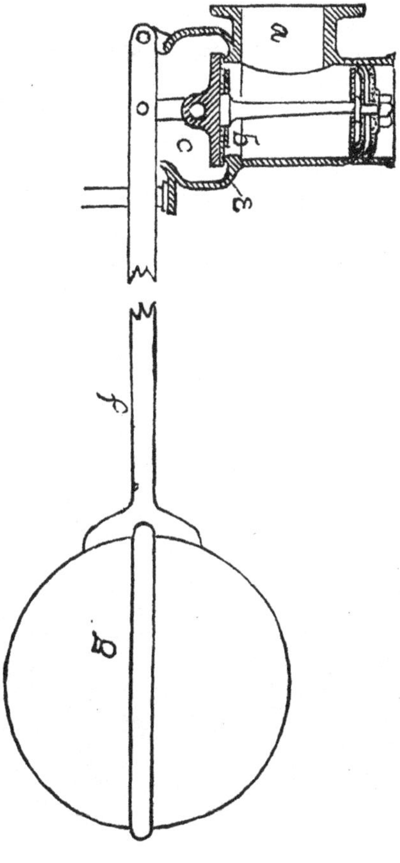

Fig. 252.

Tanks are superior to the valve as a means of flushing closets, because by their means a much larger amount of water can be thrown suddenly into the bowl of the closet, and they disconnect the closet from the main supply-pipe. Neither are they as liable to get out of order as the usual supply valve. All sanitarians agree that tanks are superior to supply-valves as a flushing arrangement for water-closets. The closet should be flushed by at least two gallons of water at each discharge of the tank. Where water is abundant the closet should receive a before-wash of a gallon and an after-wash of three gallons. With a simple closet, properly vented and connected, a simple tank properly connected, the closet all open to inspection, selected as advised in the different parts of this work, there will be no danger from sewer air being generated in or passing through the plumbing fixture which I have treated of in this book.

INDEX.

Made in the USA
Monee, IL
04 November 2026

67384992R00089